Urban and Regional Planning

University of Liverpool

This is the fifth edition of the classic text for students of urban and regional planning. It gives an historical overview of the developments and changes in the theory and practice of planning, throughout the entire twentieth century.

This extensively revised edition follows the successful format of previous editions:

- It introduces the establishment of planning as part of the public health reforms of the late nineteenth century and goes on to look at the insights of the great figures who influenced the early planning movement, leading up to the creation of the postwar planning machine.
- National and regional planning, and planning for cities and city regions, in the UK, from 1945 to 2010, is then considered. Specific reference is made to the most important British developments in recent times, including the Single Regeneration Budget, English Partnerships, the devolution of Scotland, Wales and Northern Ireland, the establishment of the Mayor of London and the dominant urban sustainability paradigm.
- Planning in Western Europe, since 1945, now incorporating new material on EU-wide issues, as well as updated country-specific sections.
- Planning in the United States, since 1945, now discussing the continuing trends of urban dispersal and social polarization, as well as initiatives in land use planning and transportation policies.
- Finally the book looks at the nature of the planning process at the start of the twenty-first century, reflecting briefly on shifts in planning paradigms since the 1960s and going on to discuss the main issues of the 1990s and 2000s, including sustainability and social exclusion and looking forward to the twenty-first century.

Sir Peter Hall is Bartlett Professor of Planning and Regeneration at the Bartlett School of Planning, University College London.

Mark Tewdwr-Jones is Professor of Spatial Planning and Governance at the Bartlett School of Planning, University College London.

Urban and Regional Planning

Fifth edition

Peter Hall
and
Mark Tewdwr-Jones

Routledge
Taylor & Francis Group

LONDON AND NEW YORK

First edition published in 1975 by David and Charles (Holdings) Ltd
Second edition published in 1985 by Unwin Hyman
Third edition published in 1992 by Routledge
Reprint 1994, 1996, 1999, 2000 (twice), 2007
Fourth edition published in 2002 by Routledge
Reprint 2007, 2008 (twice)
Fifth edition published in 2011 by Routledge
2 Park Square, Milton Park, Abingdon, Oxon, OX14 4RN

Simultaneously published in the USA and Canada
by Routledge
270 Madison Avenue, New York, NY 10016
Routledge is an imprint of the Taylor & Francis Group, an informa business

Typeset in Times New Roman and Franklin Gothic by
Florence Production Ltd, Stoodleigh, Devon
Printed and bound in Great Britain by
The MPG Books Group

British Library Cataloguing in Publication Data
A catalogue record for this book is available from the British Library

Library of Congress Cataloging in Publication Data

Hall, Peter Geoffrey.
  Urban and regional planning / Peter Hall and Mark Tewdwr-Jones. — 5th ed.
  p. cm.
  Includes bibliographical references and index.
  1. City planning. 2. Regional planning. 3. City planning—Great Britain.
  4. Regional planning—Great Britain. I. Tewdwr-Jones, Mark. II. Title.
  HT166.G7H34 2010
  361.60941—dc22   2010016606

ISBN: 978-0-415-56652-0 (hbk)
ISBN: 978-0-415-56654-4 (pbk)
ISBN: 978-0-203-86142-4 (ebk)

Contents

Plates

Figures

Tables

Preface

It is important to stress what this book is and what it is not. It is not a textbook of planning; there are excellent examples of those now available, including especially Barry Cullingworth's classic *Town and Country Planning in Britain* (14th edition, with Vincent Nadin; Routledge, 2006). Rather, it is an introduction to planning, written both for the beginning student and for the general reader. We hope that it will be found useful by students of applied geography and of town and country planning; by university and college students concerned to fit modern planning into its historical context; and by a wider audience which may want to know how planning has evolved.

Two points must be made about the treatment. First, it is deliberately historical; it traces the evolution of urban and regional problems, and of planning philosophies, techniques and legislation, from the Industrial Revolution to the present day. Second, it is necessarily written from a British standpoint for a British readership (though hopefully Commonwealth readers will find it relevant). Throughout most of the book the exclusive emphasis is on the British experience, though the survey of early planning thought in Chapter 2 is international, and Chapters 7 and 8 deliberately range out to compare the experience of other advanced industrial countries. Even in those chapters the comparison deliberately excludes the developing world; doubtless, another useful book is to be written there, but there is no space in this book to do the subject justice.

The book is a by-product of a combined 50 years of lectures on introductory applied geography and planning at the London School of Economics, the University of Reading and the University of California at Berkeley (PH); at Cardiff University and the University of Aberdeen (MTJ); and latterly given by both of us at University College London. We are grateful to successive waves of students who endured these courses and who unfailingly, by their reactions, indicated the places where material was boring or unintelligible. More particularly, we remain indebted to two academic colleagues and friends, both sadly now deceased: to the late Brian McLoughlin for his characteristically generous and thoughtful comments on a first draft; and to the late Marion Clawson for bringing his immense experience to bear on the account in Chapter 8. We must add the usual disclaimer: that for errors and omissions, the authors are solely responsible.

To two devoted helpers – Peter's secretaries, the late Monika Wheeler and Rosa Husain, who typed two successive versions of the manuscript meticulously despite unnaturally heavy departmental burdens, and the cartographer, Kathleen King – we offer our best thanks and our best wishes for Kathleen's prosperity.

Lastly, Peter would like to re-dedicate this book to his wife Magda for her imperturbable patience in the face of gross provocation in the hope that she will find the result some small recompense for many delayed dinners and obsessed weekends.

Acknowledgements

The authors and publishers would like to thank the following for granting permission to reproduce the following illustrations as listed:

California Department of Transportation, Senior Photographer Lynn G. Harrison, Plate 8.3; CORUS, Plate 5.1; The Cultural Section of the Embassy of Sweden, Plate 7.3; DATAR, Figures 7.2, 7.3; European Communities, 1995–2002, from the Eurostat website, Figure 7.1d, 7.14; Faber & Faber Ltd, Plate 3.4; Federal Office for Building and Regional Planning (BBR), Bonn, Figure 7.8; Ford Motor Company, Plate 5.2; The Head of Engineering Services, Manchester City Council, Figure 3.4; The Head of Planning Policy, Stevenage Borough Council, Figure 4.2, Plate 4.3; HMSO, Figures 6.6, 6.16; IAURIF, Figure 7.7b; KLM Aerocarto – Arnheim, Holland, Plate 7.4; The Ministry of Housing, Physical Planning and Environment, The Netherlands, Figure 7.13b; Regional Plan Association, Plates 8.1, 8.2, Figure 8.2; Simmons Aerofilms Ltd, Plates 2.2, 2.3, 2.4, 3.2, 4.1, 4.2, 5.3, 6.1, 6.2, 6.4, 6.5; John Wiley and Sons Ltd, Figure 6.15.

Every effort has been made to contact copyright holders for their permission to reprint material in this book. The publishers would be grateful to hear from any copyright holder who is not here acknowledged and will undertake to rectify any errors or omissions in future editions of this book.

1 Planning, planners and plans

Planning, the subject matter of this book, is an extremely ambiguous and difficult word to define. Planners of all kinds think that they know what it means; it refers to the work they do. The difficulty is that they do all sorts of different things, and so they mean different things by the word; planning seems to be all things to all people. We need to start by defining what exactly we are discussing.

The reference in the dictionary gives one clue to the confusion. Whether you go to the *Oxford English Dictionary* or the American *Webster's*, there you find that the noun 'plan' and the verb 'to plan' have several distinct meanings. In particular, the noun can either mean 'a physical representation of something' – as for instance a drawing or a map; or it can mean 'a method for doing something'; or 'an orderly arrangement of parts of an objective'. The first meaning, in particular, is quite different from the others: when we talk about a street 'plan' of London or New York, we mean something quite different from when we talk about our 'plan' to visit London or New York next year. But there is one definition that combines the others and blurs the distinction, as when we talk about a 'plan' for a new building. This is simultaneously a physical design of that building as it is intended to be, and a guide to realizing our intention to build it. And it is here that the real ambiguity arises.

The verb 'to plan', and the nouns 'planning' and 'planner' that are derived from it, have in fact only the second, general group of meanings: they do not refer to the art of drawing up a physical plan or design on paper. They can mean either 'to arrange the parts of', or 'to realize the achievement of', or, more vaguely, 'to intend'. The most common meaning of 'planning' involves both the first two of these elements: planning is concerned with deliberately achieving some objective, and it proceeds by assembling actions into some orderly sequence. One dictionary definition, in fact, refers to what planning does; the other, to how planning does it.

The trouble arises because, although people realize that planning has this more general meaning, they tend to remember the idea of the plan as a physical representation or design. Thus they imagine that planning must include the preparation of such a design. Now it is true that many types of planning might require a physical design, or might benefit from having one: planning is often used in the production of physical objects, such as cars or aeroplanes or buildings or whole towns, and in these cases a blueprint of the desired product will certainly be needed. But many other types of planning, though they will almost certainly require the production of many symbols on pieces of paper, in the form of words or diagrams, may never involve the production of a single exact physical representation of the entity which is being produced.

For instance, the word 'planning' is today applied to many different human activities – in fact, virtually all human activities. One almost certainly needs a plan to make war; diplomats make contingency plans to keep the peace. We talk about educational planning: that does not mean that every detail of every class has to be planned by some bureaucracy (as happens, by repute, in France), but mercly that advance planning is necessary if students are to find classrooms and libraries and teachers when they arrive at a certain age and seek a certain sort of education. We talk about planning the economy to minimize the swings of boom and slump, and to reduce the misery of unemployment; we hear about a housing plan and a social-services plan. Industry now plans on a colossal scale: the production of a new model of a car, or a personal computer, has to be worked out long in advance of its appearance in the shops. And all this is true, whatever the nature of the economic system. Whether labelled free enterprise or social-democratic or socialist, no society on earth today provides goods and services for its people, or schools and colleges for its children, without planning. One might regret it and wish for a simpler age when perhaps things happened without forethought; if that age ever existed, it has gone for ever.

The reason is the fact of life everybody knows: that modern society is immeasurably more complex, technically and socially, than previous societies. Centuries ago, when education involved the simple repetition of a few well-understood rules which were taught to all, and when books were non-existent, the setting up of a school did not involve much elaborate plant or the training of specialized teachers. The stages of production were simpler; wood was cut in the forest, people wrought it locally into tools, the tools were used by their neighbours, all without much forethought. But today, without elaborate planning, the complex fabric of our material civilization would begin to crack up: supplies of foodstuffs would disappear, essential water and power supplies would fail, epidemics would rapidly break out. We see these things happening all too readily, after natural or human disasters like earthquakes or wars or major strikes of railway- or power-workers. Though some of us may decide to opt out of technological civilization for a few years or for good, the prospect does not seem likely to appeal to the great mass of humankind even in the affluent world. Those in the less affluent world are in much less doubt that they want the security and dignity that planning can bring.

The point is that the sorts of planning which we have been discussing in these last two paragraphs either may not require physical plans at all, in the sense of scale blueprints of physical objects, or may require them only occasionally or incidentally. Planning is more likely to consist, for the most part, of written statements accompanied by tables of figures, or mathematical formulae, or diagrams, or all these things. The emphasis throughout is on tracing an orderly sequence of events which will achieve a predetermined goal.

Consider educational planning as an example. First, the goal has to be fixed. It may be given externally, as a situation which has to be met: to provide education which will meet the expected demands ten years hence. Or there may be a more positive, active goal: to double the numbers of scientists graduating from the universities, for instance. Whatever the aim, the first step will be a careful projection which leads from the present to the future target date, year by year. It will show the number of students in schools and colleges and the courses that will be needed to meet whatever objective is stated. From this, the implications will be traced in terms of buildings, teachers and materials. There may need to be a crash school-building programme using quickly assembled prefabricated components, a new or a supplementary teacher-training programme or an attempt to win back married women into teaching, or a new series of textbooks or experiments in web-based learning, all of which in turn will take time to set in motion and produce results. At critical points in the process, alternatives will be faced. Would it be more economical, or more effective, to increase teacher supply or concentrate on a greater supply of teaching material through the Internet? Could better use be made of existing buildings by better overall coordination, rather than by putting up new buildings? Ways will need to be found of evaluating these choices. Then, throughout the lifetime of the programme, ways will need to be found of monitoring progress very closely to take account of unexpected failures or divergences from the plan or changes in the situation. In the whole of this complex sequence the only scale models may be the designs of the new schools or of the IT system and a few other details – a small part of the whole, and one which comes at a late stage in the process, when the broad outlines of the programme are determined.

To summarize, then: planning as a general activity is the making of an orderly sequence of action that will lead to the achievement of a stated goal or goals. Its main techniques will be written statements, supplemented as appropriate by statistical projections, mathematical representations, quantified evaluations and diagrams illustrating relationships between different parts of the plan. It may, but need not necessarily, include exact physical blueprints of objects.

The application to urban and regional planning

The difficulty now comes when we try to apply this description to the particular sort of planning that is the subject matter of this book: urban and regional planning (or, as it is often still called, town and country planning). In many advanced industrial countries, such as Britain, the United States, Germany or Japan, the phrase 'urban planning', or 'town planning', is strictly a tautology: since a great majority of the population are classed in the statistics as urban and live in places defined as urban, 'town planning' seems simply to mean any sort of planning whatsoever. In fact, as is well known, 'urban planning' conventionally means something more limited and precise: it refers to planning with a spatial, or geographical, component, in which the general objective is to provide for a spatial structure of activities (or of land uses) which in some way is better than the pattern existing without planning. Such planning is also known as 'physical planning'; 'spatial planning' is perhaps a more neutral and more precise term.

If such planning centrally has a spatial component, then clearly it only makes sense if it culminates in a spatial representation. Whether this is a very precise and detailed map, or the most general diagram, it is to some degree a 'plan' in the first, more precise meaning of the term. In other words, it seems that urban planning (or regional planning) is a special case of general planning, which does include the plan-making, or representational, component.

Broadly, in practice this does prove to be the case. It is simply impossible to think of this type of planning without some spatial representation – without a map, in other words. And whatever the precise organizational sequence of such planning, in practice it does tend to proceed from very general (and rather diagrammatic) maps to very precise ones, or blueprints. For the final output of such a process is the act of physical development (or, in some cases, the decision not to develop, but to leave the land as it is), and physical development, in the form of buildings, will require an exact design.

A great deal of discussion and controversy in recent years tends to have obscured this fact. In most countries spatial or urban planning as practised for many years – both before the Second World War and after it – was very minute and detailed: the output tended to consist of very precise large-scale maps showing the exact disposition of all land uses and activities and proposed developments. During the 1960s such detailed plans were much attacked: planning, it was argued, needed to concentrate much more on the broad principles rather than on details; it should stress the process, or time sequence, by which the goal was to be reached, rather than present the desired end-state in detail; it should start from a highly generalized and diagrammatic picture of the spatial distributions at any point of time, only filling in the details as they needed to be filled in, bit by bit. This, as we shall see later, is the essential difference in Britain between the system of local town and country planning introduced by the historic Town and Country Planning Act of 1947, and the system which replaced it under the Town and Country Planning Act of 1968.

The central point, though, is that this type of planning is still essentially spatial – whatever the scale and whatever the sequence. It is concerned with the spatial impact of many different kinds of problem, and with the spatial coordination of many different policies. Economic planners, for instance, are concerned with the broad progress of the economy, usually at national and sometimes at international level: they look at the evolving structure of the economy, in terms of industries and occupations, at the combination of the factors of production which brings forth the flow of goods and services, at the income thus generated and its reconversion into factors of production, and at problems of exchange. Regional economic planners will look at the same things, but always from the point of view of their particular spatial impact: they consider the effect of the variable, geographical space and distance, on these phenomena. Similarly, social planners will be concerned with the needs of the individual and the group; they will be concerned with the changing social structure of the population, with occupational mobility and its effect on lifestyles and housing patterns, with household and family structure in relation to factors like age and occupation and educational background with household income and its variation, with social and psychological factors which lead to individual or family breakdown. The social planner in the urban planning office shares the same interests and concerns, but sees them always with the spatial component: s/he is concerned, for instance, with the effect of occupational mobility on the inner city – as against the new suburb – on changing household structure as it affects the housing market near the centre of the city, and on household income in relation to items like travel cost for the low-income family whose available employment may be migrating to the suburbs.

The relationship between urban and regional planning and the various types of specialized planning, in these examples, is interestingly like the relationship of geography, as an academic subject, to other related social sciences. For geography also has a number of different faces, each of which stresses the spatial relationship in one of these related sciences: economic geography analyses the effect of geographic space and distance on the mechanisms of production, consumption and exchange; social geography similarly examines the spatial impact upon patterns of social relationship; political geography looks at the effect of location upon political actions. One can argue from this that spatial planning, or urban and regional planning, is essentially human geography in these various aspects, harnessed or applied to the positive task of action to achieve a specific objective.

Many teachers in planning schools would hotly deny this. They would argue that planning, as they teach it, necessarily includes many aspects which are not commonly taught in geography curricula – even those that stress the applications of the subject. The law relating to the land is one of these; civil engineering is another; civic design is another. This is true, though many would argue – both inside the planning schools, and out – that not all these elements are necessary to the planning curriculum. What does seem true is that the central body of social sciences which relate to geography, and whose spatial aspects are taught as parts of human geography – economics, sociology, politics and psychology – does form the core of the subject matter of urban and regional planning. By 'subject matter' we mean that which is actually planned. It is, however, arguable that there is another important element in planning education, not covered in this body of social science: that is, the study of the process of planning itself – the way we assume control over physical and human matter, and process it to serve their defined ends. According to this distinction, 'planning method' would be what is common to the education of all kinds of planners – whether educational, industrial, military or any other; geography and its related social sciences would constitute the peculiar subject matter of that particular division of planning called urban and regional.

'Planning' as an activity

What then would this core of planning education – the study of planning process – comprise? This is a basic question which ought to have been the subject of intense debate in schools of planning. But curiously, for a long time it was avoided – the reason being, apparently, that planning education was seen as education in making physical plans, not education in planning method. The first people to raise the question seriously were not teachers of physical planning, but teachers of industrial or corporate planning, in the American business schools. There, up to about 1945, education in management was usually based on a rather narrow spectrum of skills in applied engineering and accounting; the aim was to obtain maximum efficiency in plant operation, both in an engineering sense and in an accounting sense, and little attention was given to the problems of decision-taking in complex situations. But during the 1950s, partly as the result of the work of such fundamental thinkers as Chester Barnard, Peter Drucker and Herbert Simon, management education was transformed. First, it developed into a science of decision-making, which borrowed freely from concepts in philosophy and politics, and second, it harnessed the thinking of a number of social sciences, such as economics, sociology and psychology. It was this new tradition in corporate planning which began, after about 1960, to affect the direction and content of education for physical planning.

By this time, however, management education had further evolved. With the development of computerization in management and planning of all kinds, there was increasing interest in the development of sophisticated control systems which would automatically control machinery. Such systems, of course, were only a development of earlier experiments in automation, which can be dated right back to the origins of the Industrial Revolution; but progress in this field took a big leap forward with the rapid development of more complex computers during the 1950s. Yet even before this, a remarkable original thinker, Norbert Wiener of Harvard, had anticipated the development and much more. In a book published in 1950, *The Human Use of Human Beings*, he had suggested that automation would liberate the human race from the necessity to do mundane tasks. But further, he proposed that the study of automatic control systems was only part of a much larger science of cybernetics,[1] which he defined in the title of a book published in 1948 as the science of 'Control and Communication in the Animal and the Machine'. According to Wiener, animals and especially human beings have long possessed extremely complex communication and control mechanisms – the sort of thing the computer was then replicating. Human societies, Wiener suggested, could be regarded as another manifestation of this need for communication and control.

Thus a new science was born. Rapidly developing in the late 1950s and 1960s, it had a profound influence on research and education in management, and particularly in planning. For if human arrangements could be regarded as complex interrelating systems, they could be paralleled by similar systems of control in the computer, which could then be used to monitor developments and apply appropriate adjustments.

The best analogy, much quoted at that time, was manned space flight. In an expedition to the moon most of the adjustments to the spacecraft were made not by the astronauts but by an extraordinarily complex computer-control system on earth at Houston, Texas. Similarly, it is argued, the development of cities and regions could be controlled by a computer which received information about the course of development in a particular area, related this to the objectives which had been laid down by the planners for the development during the next few years and thus produced an appropriate series of adjustments to put the city or the region 'on course' again.

In practice this insight has been very useful for the way we think about the physical or spatial planning. Information systems are now used very widely in the planning process. And, as we shall see in later chapters of this book, it has profoundly affected the way planners think about their job and the way they produce plans. In essence it has led to a swing away from the old idea of planning as production of blueprints for the future desired state of the area, and towards the new idea of planning as a continuous series of controls over the development of the area, aided by devices which seek to model or simulate the process of development so that this control can be applied. This in turn has led to a complete change in the sequence of the planners' work. Formerly, at any time from about 1920 until 1960, the classic sequence taught to all planning students was survey–analysis–plan. (The notion of survey before plan had first been worked out, and taught, by a remarkable British pioneer in planning, Patrick Geddes; his work is discussed in more detail in Chapter 3.) The terms were self-explanatory. First, the planner made a survey, in which s/he collected all the relevant information about the development of his or her city or region. Then s/he analysed these data, seeking to project them as far as possible into the future to discover how the area was changing and developing. And third, s/he planned: that is, s/he made plans which took into account the facts and interpretations revealed in the survey and analysis, and which sought to harness and control the trends according to principles of sound planning. After a few years – the British Planning Act of 1947 laid down that the period should be every five years – the process should be repeated: the survey should be carried out again to check for new facts and developments, the analysis should be reworked to see how far the projections needed modifying and the plan should be updated accordingly.

The new planning sequence, which has replaced this older one as orthodoxy, reflects the approach of cybernated planning. It is more difficult to represent in words because it is a continuous cycle; more commonly, it is represented as a flow diagram. But, to break into the flow for purposes of exposition, it can be said to start with the formulation of goals and objectives for the development of the area concerned. (These should be continuously refined and redefined during the cycles of the planning process.) Against this background the planner develops an information system which is continuously updated as the region develops and changes. It will be used to produce various alternative projections, or simulations, of the state of the region at various future dates, assuming the application of various policies. (The aim is always to make this process as flexible and as varied as possible, so that it is possible to look at all sorts of ways of allowing the region to grow and change.) Then the alternatives are compared or evaluated against yardsticks derived from the goals and objectives, to produce a recommended system of policy controls which in turn will be modified as the objectives are re-examined and as the information system produces evidence of new developments. Though it is difficult to put this new sequence into a string of words like the older one, it might be succinctly described as goals–continuous information–projection and simulation of alternative futures–evaluation–choice–continuous monitoring. Something like this sequence, with some differences in words and in ordering, can be found in several important and well-known accounts of the planning process written in the 1960s and early 1970s.

Objectives in planning – simple and complex

In practice, as has been said above, this is a great improvement. It means that the whole planning process is more clearly articulated, more logical and more explicit. It is obviously better that planners should start with a fairly exhaustive discussion about what they are seeking to achieve and that they should go on having this discussion during the whole planning process. It is better, too, that different alternatives for the future should be developed, so that they can be openly discussed and evaluated. And the emphasis on specific evaluation, using certain fixed criteria, is an advance. Planning is now much more flexible, working with much greater information. And it is more rational – at least potentially so.

Nevertheless, the alternative system has proved to create many new problems and pitfalls of its own. The development of computerization does not make planning easier, in the sense that it somehow becomes more automatic. There may be many automatic aids to smooth out tedious processes, such as detailed calculations; but they do not diminish the area of human responsibility – the responsibility to take decisions. And the basic difficulty is that it is more difficult, and finally less feasible, to apply cybernation to most urban planning problems than it is to apply it to the job of getting human beings on the moon.

At first sight this may seem absurd: nothing could be more complex than space travel. But this is to mix up levels of complexity. Space travel (or, indeed, commercial aviation) presents many technical problems, but there are two features that make it basically simple. First, the objective is clearly understood. Second, the processes involved are nearly all physical: they are subject to laws of physics, which are much better understood, and which appear to be more regular in their application, than laws of human behaviour. (There are human beings involved, of course, but in practice they are reduced to little more than biological units for most of the voyage.) The kind of planning that most resembles space travel is transportation planning, and it is significant that this was where computerized systems planning had its earliest and most successful applications.

Elsewhere, it has proved harder. That is because it is inherently more complex. First, the basic objective is not well understood; there is clearly more than one objective, and perhaps dozens (economic growth, fair distribution of income, social cohesion and stability, reduction of psychological stress, a beautiful environment – the list seems endless). These objectives may not be readily compatible, and may indeed be contradictory. Second, most of the processes which need controlling are human processes, which are less well understood and work with much less certainty than laws in the physical sciences. Anyone who has studied any of the social sciences such as economics, sociology, psychology or human geography is familiar with this fact. Just as in these sciences we have to work with laws of statistical tendency rather than with laws which are constantly reliable in producing experimental results, so it will be in much of spatial or physical planning.

One point made in the last paragraph is relevant for our understanding of the particular nature of spatial planning. Earlier, we said that its method was shared with other sorts of planning activity; its subject matter was distinctively spatial, so that at some time, in some sense, it would produce spatial representations of how activities should be ordered on the ground. We now see that spatial planning, as we are using the term in this book – urban and regional planning, as it is conventionally termed – has another feature: it is multidimensional and multi-objective planning. It is necessary to specify these two linked attributes, because there are many types of planning which are 'spatial' in the sense that they are concerned with spatial arrangements on the earth's surface, but have only a single dimension and a single objective. When sanitary engineers consider a sewer plan, their work certainly has a spatial component, but it is neither multidimensional nor multi-objective. (Or, to be more precise, even if the engineer thinks s/he has more than one objective, these are all engineering objectives within the same basic dimension.)

This engineer, or colleagues like the highway engineer or telephone engineer, are doubtless all working with plans which are spatial representations of their territory. But none of them will be trying (for instance) to balance the advantages of preserving a long-established inner-city society against the advantages of building better housing on an estate some distance away, or the problem of reconciling higher car ownership with the preservation of public transport for those who have no access to cars and the preservation of a decent urban environment, or the merits of segregating factory zones versus the merit of having local factories nearer to people's homes – all of these, and many more, being considered as part of the same planning process, and having finally each to be considered *vis-à-vis* all the others. This task of reconciliation is the essence of the job of the urban and regional planner; this is why, compared with most other sorts of job regarded as planning, it is so difficult.

It is difficult in two ways. First, the amount of necessary information and specialized expertise is so much greater than in most other planning activities: it covers almost the whole of human experience. The ideal urban and regional planner would have to be a good economist, sociologist, geographer and social psychologist in his or her own right, as well as having several other necessary physical-scientific skills, such as a good understanding of civil engineering and of cybernetics. To judge the quality of the information s/he was receiving, s/he would need to be a sophisticated (and even slightly sceptical) statistician. And s/he would need to be a highly competent systems analyst in order to develop the relationships with the computer control system with which s/he related. All of which, of course, constitutes an impossible specification – and a daunting task for the educationalist.

But second, and even more problematically, there is the need to frame and then weigh up different objectives. Consider a very typical (and very topical) type of planning controversy: the line of a new urban motorway. Some critics say that it would be quite unnecessary if public transport were adequate: some that the line should be shifted. The fact is that car ownership is rising, and this seems outside the planner's control; it is set by the political or social framework within which s/he acts. The projections (which may not be entirely reliable) suggest that the traffic will overwhelm the present road network, giving an environment to many thousands of people which, by current standards, is judged intolerable. The quality of public transport is declining, but the available evidence shows that better quality would not have much result in tempting people back from their cars and reducing the case for the motorway. One possible line for the motorway goes through a slum district due for early demolition and rebuilding; some sociologists say that the community should be rehoused *in situ*, others argue that many of the people would lead happier lives in a new town. Another line goes through open space which contains playing fields as well as the nesting grounds of several species of birds; local sports clubs and nature conservationists are united in opposing this line. The costs to the public purse are known in the two cases, but the benefits are dependent on the valuation of travel time for the likely motorway users, on which two groups of economists are hotly disputing. And the costs, or disbenefits, for different groups of the public affected by the building of the motorway are almost incalculable.

There are many varying interests and special academic skills, some of the practitioners of which cannot agree among themselves; the only person who seems competent to take any decision at all is someone whose training and thinking are supposed to encompass them all. This, of course is the general urban and regional planner. This is not the point at which to discuss the resolution of the problem just mentioned; in fact, there simply is no clear resolution, and the most the planner can do is to try to reach a decision within a clear and explicit framework – which, hopefully, the present style of planning helps him or her to do.

The example has been given simply to illustrate the unique quality, and the unique difficulty, of the sort of planning that is the subject matter of this book. To sum up: urban and regional planning is spatial or physical: it uses the general methods of planning to produce a physical design. Because of the increasing influence of these general methods, it is oriented towards process rather than towards the production of one-shot (or end-state) plans. Its subject matter is really that part of geography which is concerned with urban and regional systems; but the planning itself is a type of management for very complex systems. And further, it is necessarily multidimensional and multi-objective in its scope; this is what distinguishes it from the work of many other professionals whose work can fairly be described as planning with a spatial component.

Structure of this book

The remainder of this book falls into five parts. Chapters 2 and 3 outline the early history of urban development in Britain, with special references to the changes brought about by the Industrial Revolution, and the contributions of notable early thinkers and writers on urban planning during the period 1880–1945. Chapter 4 takes the British story through the 1930s and 1940s, describing the new challenge of regional imbalance which appeared in the Great Depression of 1929–32, and the subsequent creation of the postwar planning machine following publication of the Barlow Report of 1940. Chapters 5 and 6 analyse the postwar history, and attempt to pass judgement on the performance of the planning system, first at broad regional level in respect of economic planning, then at the scale of the town and the city region in respect of urban planning. Chapters 7 and 8 attempt a comparative look at planning experience in other developed industrial countries, Chapter 7 for Western Europe and Chapter 8 for the United States. Lastly, Chapter 9 provides an outline of the sequence of urban and regional plan-making, with an introduction to some of the more important techniques involved at various stages of this process; it is deliberately written to provide a bridge to the more advanced textbooks of planning, which deal with these processes in more detail. But this book, as we have stressed in the Preface, must end there; it does not try to compete with those textbooks, but to provide the necessary historical framework of introduction to them.

Further reading

Andreas Faludi, *Planning Theory* (Pergamon, 1973), is a good introduction to these questions. John Friedmann, *Planning in the Public Domain* (Princeton University Press, 1988), is the most comprehensive treatment. Good recent accounts are from Patsy Healey, *Land Use Planning and the Mediation of Urban Change: The British Planning System in Practice* (Cambridge University Press, 1988) and *Collaborative Planning: Shaping Places in Fragmented Societies* (Macmillan, 1997).

Note

1   The word is derived from an ancient Greek word, meaning 'helmsman' or 'oarsman'.

2 The origins: urban growth from 1800 to 1940

Modern urban and regional planning has arisen in response to specific social and economic problems, which in turn were triggered by the Industrial Revolution at the end of the eighteenth century. It is important to notice that these problems did not all come at once, in the same form; they changed in character, and in their relative importance, so that the questions uppermost in the minds of city-dwellers in the 1930s were by no means the same as those experienced by their great-grandfathers in the 1840s. As problems were identified, solutions were proposed for them; but because of the inertia of people's minds, and still more the inertia of social and political processes, these solutions – especially the more radical ones – might not be put into action until decades afterwards, when the problem itself had changed in character and perhaps also in importance. That is a most important common theme which runs through this and the next two chapters.

Planning before the Industrial Revolution

There were important cities before the Industrial Revolution: Ancient Rome had an estimated population of one million by the year AD 100; Elizabethan London numbered about 200,000 people. Correspondingly, these cities had problems of economic and social organization: Rome had to be supplied with water brought over considerable distances by aqueduct (the word itself is Roman in origin), and the city developed immense problems of traffic congestion – which, unfortunately, have been inherited by the modern city 2,000 years later. London by the fourteenth century had to draw on coalfields by the River Tyne, 270 miles away, for fuel, and on distant countries for more specialized provisions, such as dyestuffs or spices; by the seventeenth century it, too, was drawing water from 35 miles away by aqueduct. (The New River, which runs through North London, is part of it.) These problems in turn brought forth a host of regulations for the better ordering of the city, sometimes dealing with strangely modern problems: Rome banned chariots at night to deal with the first recorded case of urban noise pollution; in London in the fourteenth century a man was hanged for burning 'sea coal' – a somewhat draconian penalty for medieval air pollution.

Furthermore, many cities both in the ancient and the medieval world were planned, at least in the sense that their existence and their location were laid down consciously by some ruler or some group of merchants; and among this group, a large proportion even had formal ground plans with a strong element of geometric regularity. In Britain the group of medieval planned towns is larger than many people think: a small town like Baldock, on the Great North Road (A1) before it was bypassed, was actually a creation of the Knights Templar, and the name itself is a corruption of Baghdad; Winchelsea on the Sussex coast, and small towns in North Wales like Flint, Conway and Caernarvon, were all fortified towns created by Edward I in the late thirteenth century, and were deliberately modelled on the Bastide towns established by the French kings as part of their conquest of Provence a few years earlier.

The greatest flowering of formal town planning before the Industrial Revolution, though, came in what is known in Continental Europe as the Baroque era: the seventeenth and eighteenth centuries. There, it produced such masterpieces of large-scale architectural design as the reconstruction of Rome during the late sixteenth and early seventeenth centuries; or the great compositions of the Tuileries gardens and the Champs-Elysées, in Paris; or the palace of Versailles and its bordering planned town; or the completely planned town of Karlsruhe, in Germany; or the seventeenth-century quarters of Nancy, in the province of Lorraine in eastern France; as well as many other smaller, but fine, examples. These were nearly all expressions of absolute regal or papal power, and some commentators have claimed to see in them the expression of a new style of warfare – instead of the medieval walled town, cities must now be planned along broad formal avenues where mobile armies could deploy themselves. Britain, after Cromwellian times, had no such absolute monarchy; here the aristocracy and the new merchant class dominated the growth of cities, and determined their form. The result was a different but equally distinctive form of town planning: the development of formal residential quarters consisting of dignified houses built in terraces or rows, generally on a strongly geometrical street plan which was modified by charming squares with gardens. The original development of many of the quarters of London's West End, now sadly decimated by later reconstruction – areas like St. James's, Mayfair, Marylebone and Bloomsbury – still provides the best examples in Britain of this type of planning attached to an existing major city; Edinburgh's New Town, facing the medieval city across the deep cut now occupied by the railway, is another. But perhaps the best example of eighteenth-century British town planning is the development of Bath, up to then a small medieval town, as the result of a new enthusiasm for spa cures among the aristocracy at that time.

All these examples, and many other imitations, have great interest for the student of architecture or the origins of planning. And similarly, the creation of the rural landscape of Europe – a process which involved much more conscious planning than most people, looking at the result casually, would imagine – is important for the planner, understanding how previous generations adjusted to the opportunities and the limitations the region presented. But the subject deserves much fuller treatment than it can receive here; and it is excellently written up in the book *The Making of the English Landscape* by W.G. Hoskins. Our main concern now is a subject that has little relation with the past: the unprecedented impact of modern industrialism on urban development and upon consequent urban planning problems.

The impact of industrialism

Oddly, at first the Industrial Revolution had no striking effect on urban growth. The earliest of the new inventions in textiles or in iron-making, developed in England between 1700 and 1780, seemed rather to be dispersing industry out of the towns and into the open countryside. By the end of this period – and even later still, into the early nineteenth century – typical industrial landscapes, such as the cotton-making areas of south Lancashire or south Derbyshire, the woollen areas of the Colne and Calder Valleys in the West Riding of Yorkshire, or the iron-making and working areas of Coalbrookdale (Shropshire) or the Black Country, essentially consisted of a straggle of small industrial hamlets across an area that was fundamentally still rural. In some industries this tradition survived even longer: D.H. Lawrence's early novels describe it in the Nottinghamshire coalfields as late as 1900.

But it was coal that changed the situation. As soon as it became a principal raw material of industry – replacing water power in textiles after 1780, for instance – it tended to concentrate industry where supplies could be made available: on the coalfields themselves, and then adjacent to bulk transport. Britain, because it industrialized earlier than any other country, experienced special constraints on its industrial location pattern: the early machinery consumed great quantities of coal because it was inefficient, and the coal was very expensive to transport because there were no railways, only canals. After about 1830 (the first steam-driven railway, the Stockton and Darlington, came in 1825) both these conditions changed, and industry was freer to locate. But by then its pattern was fixed (Figure 2.1).

This fact alone created a new phenomenon: the new industrial town, developed almost from nothing – or perhaps from a small and obscure village origin – within a few years, on the coalfields of Lancashire and Yorkshire and Durham and Staffordshire. Simultaneously, some towns – those which were neither port towns nor on coalfields – stagnated industrially. But many older-established medieval towns – because they were near enough to coalfields, or because they were on navigable water, or because they became railway junctions soon after the railways arrived – were also able to become major centres of the new factory industry: Leicester, Nottingham and Bristol are good examples. Port towns, indeed, were just as important as pure industrial towns in the whole process of industrialization, because they effected the critical exchange of raw materials and finished products on which the whole system depended: thus cities like Liverpool, Hull, Glasgow and, above all, London were among the fastest-growing places from 1780 onwards.

Some of the resulting growth patterns are extraordinary – even by the standards of the twentieth century, which has become used to mushrooming growth in the cities of the developing world. The most spectacular cases were, of course, some of the new industrial towns which developed almost from nothing. Rochdale in Lancashire, for instance, numbered about 15,000 in 1801, 44,000 in 1851 and 83,000 in 1901; West Hartlepool in County Durham grew from 4,000 in 1851 to 63,000 by 1901. But though their percentage rate of growth was necessarily more modest, many bigger and older-established centres managed to maintain an amazing rate of growth throughout most of the century. London doubled from approximately 1 million to about 2 million between 1801 and 1851; doubled again to 4 million by 1881; and then added another 2.5 million to reach 6.5 million in 1911.

The parallel with the cities of the developing world is, in several ways, only too exact. The people who flooded into the burgeoning nineteenth-century industrial and port cities of Britain were overwhelmingly coming from the countryside. They tended to be drawn from the poorer section of the rural population: those who had least to lose and most to gain by coming to the city. Many of them had found it increasingly difficult to get work after the enclosure movement, which, approved and planned by Parliament, transformed so much of midland and southern England during the eighteenth century. Some of them – like the Irish who flooded into Liverpool and Manchester and Glasgow after the failure of the potato harvest in 1845–6 – were truly destitute. They had little or no knowledge of the technical skills needed by the new industry, or of the social and technical necessities of urban life. And though the industry of the towns provided economic opportunities in plenty for an unskilled labour force, the social arrangements in the towns were quite incapable of meeting their needs for shelter, for elementary public services like water and waste disposal, or for health treatment.

This last point is critical: these towns had only the most elementary arrangements, or none, for providing water, or clearing refuse or sewage, or for treating mass epidemics.

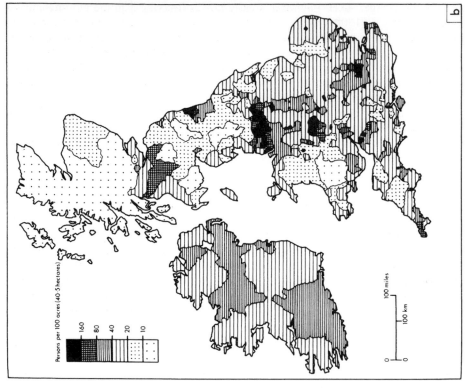

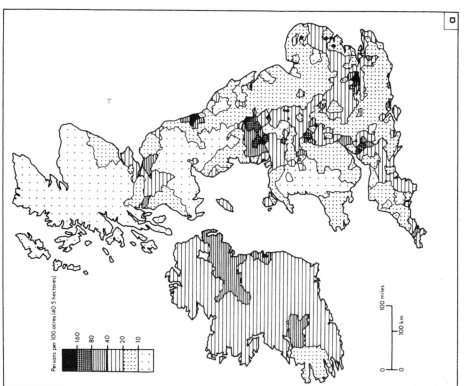

Figure 2.1 Population distribution in the British Isles: (a) 1801 (Ireland, 1821); (b) 1851 (Ireland, 1841). In the first half of the nineteenth century population concentrated in the towns – especially on the newly developed coalfields of the north. Here, towns grew without plan or control.

Many of the towns, having sprung up so rapidly from villages, had virtually no arrangements at all. Even in the larger towns they had been very elementary; and they tended to be quite overwhelmed by the influx. In a stagnant or slowly growing city, or in a relatively small town, the consequences might not have been so dire: wells might not have become polluted so easily by sewage; new dwellings could be constructed quite easily outside the existing town limits without overcrowding. But in the rapidly growing towns, these solutions were not open. Because there was no system of public transport to speak of, the new population like the old must be within walking distance of work in the factories or warehouses. Within the limits thus set, population densities actually tended to rise during the first half of the nineteenth century; the census records for London or Manchester show this quite clearly.

The results could have been predicted. Limited water supplies were increasingly contaminated by sewage; there were quite inadequate arrangements for disposal of waste, and filthy matter of all kinds remained close to dense concentrations of people; water supplies were lacking or fitful, and personal hygiene was very poor; overcrowding grew steadily worse, both in the form of more dwellings per acre and more people per room; cellar dwellings became all too common in some cities, such as Manchester or Liverpool; medical treatment, and above all public health controls, were almost completely lacking. And, to make things worse, the greater mobility induced by trade meant that epidemics could move more rapidly across the world than ever before. This, plus polluted water supplies, was the basic cause of the terrible cholera epidemics that swept Britain in 1832, 1848 and 1866.

The results for public health became clear only after the establishment of an efficient government organization for charting the state of public health: the General Register Office, set up in 1837. William Farr, the first Registrar General and one of the founding fathers of the modern science of statistics, showed as early as 1841 that the expectation of life at birth – 41 years in England and Wales overall and 45 in salubrious Surrey – was only 26 years in Liverpool: two years later in Manchester, it was only 24. Much of this difference arose because of the shockingly high infant-mortality rates in the northern industrial towns: 259 out of every 1,000 children born died within the first year of life in Liverpool in 1840–1, and for the early 1870s the average was still 219. (The corresponding figure for Liverpool in 1970 was 21; an eloquent testimony to the improved quality of life.)

It was a situation which society could not tolerate for long. Even in the most cynical view, the more privileged members of society – the industrialists and merchants – were likely to suffer from it: less so than their workers, to be sure, but still the statistics showed that the risk was considerable. Yet the struggle for reform was a difficult one; it had to surmount at least three major hurdles. The first was the will to act; and this took some time to spread to a large section of the controlling interests who dominated Parliament and the local authorities. Until 1832, it must be remembered, Parliament was totally unrepresentative of the experience and the views of those in the industrial towns. The second was knowledge of how to act; and on many critical questions, such as the germ-borne causation of disease and its treatment, and above all the origin of cholera, medical experts were sadly ignored until after the turn of the century. (Cholera was first identified as a water-borne disease in 1854 by a London doctor, John Snow, in an early piece of spatial analysis (Figure 2.2); he proved that the outbreaks in a slum district of London were systematically associated with the water supply from a single pump. But it was not until some time after this that the mechanism was understood.) The third need was for effective administrative machinery, including finance, for instituting the necessary controls and providing public services; and this, in an era of rampant *laissez-faire*, was in many ways the hardest of all, involving as it did the reform of existing and ineffectual local governments.

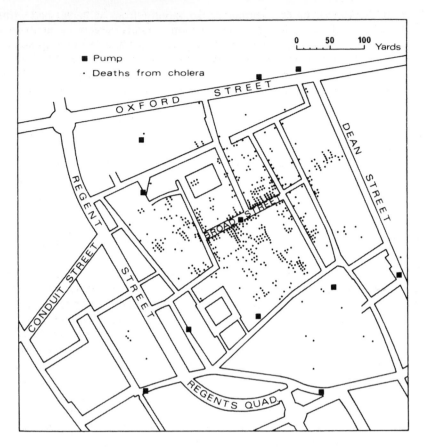

Figure 2.2 Deaths from cholera in the Soho district of London, September 1854. Dr John Snow's celebrated map, which established the connection between the cholera outbreak and a single polluted water pump in Broad Street. This emphasized the importance of supplying pure water to the inhabitants of the growing cities in Britain.

The position here was confused. An Act of 1835, the Municipal Corporations Act, had reformed borough government; but usually it did not make the boroughs exclusively responsible for public services or sanitary controls, and in any case, many towns had new building beyond their boundaries. Two major Blue Books, or official reports – the Select Committee on the Health of Towns (1840) and the Royal Commission on the State of Large Towns (1844–5) – recommended that there should be a single public health authority in each local area to regulate drainage, paving, cleansing and water supply; they also called for powers to govern the standards of construction of new buildings. From the mid-century a series of Acts – the Public Health Act of 1848, which set up a Central Board of Health and allowed it to establish Local Boards of Health, the Nuisance Removal Acts from 1855 and the Sanitary Act of 1866 – aided the control of the more obvious sanitary problems. And, from the 1860s, there was increasing interest in the control of building standards. The Torrens Acts from 1868 onwards allowed local authorities to compel owners of insanitary dwellings to demolish or repair them at their own expense; the Cross Acts, from 1875 onwards, allowed local authorities themselves to prepare improvement schemes for slum areas (Plate 2.1). The last of these Acts – the Public Health Act of 1875 – produced a long-overdue fundamental reform of local government in England and Wales, outside the boroughs: the country was divided into urban and rural sanitary districts, which would be supervised by a central government department already set up in 1871, the Local Government Board. The word 'sanitary' in their title amply indicates the original scope of these authorities; but soon they were incorporated in a comprehensive local government reform. Three Acts – the Municipal Corporations Act of 1882 and the Local Government Acts of 1888 and 1894 – gave new local government structures to the boroughs, the counties (plus new county boroughs or large towns) and county districts. This system survived almost unchanged until the major reform of English local government, carried through by the Act of 1972.

These local authorities, but above all the boroughs, increasingly began to adopt model by-laws for the construction of new housing from the 1870s onwards. By-law housing, as it came to be known, can readily be recognized in any large British city. It tends to occur in a wide ring around the slums of the earlier period (1830–70), most of which were swept away in the great assault on the slums between 1955 and 1970. Drably functional, it consists of uniform terraces or rows of two-storey housing in the local building material (brick in most parts of the country, stone on the upland borders in Lancashire and the West Riding). The streets have a uniform minimum width to guarantee a modicum of air and light; each house originally had a separate external lavatory with access to a back alley, which runs parallel to the street. (This was necessary for the emptying of earth closets; for even in the 1870s it was impossible to provide for water-borne clearance of waste from many of these houses. It was also thought desirable for clearance of solid refuse.) So originally, these houses had neither inside lavatory nor inside bath – indeed, no fixed bath at all. They represented a housing problem in the 1960s and 1970s; but since they were built according to some minimal standards, most have been upgraded to reasonable modern standards without the need for demolition. And the areas in which they were concentrated were at the same time often environmentally improved, enhancing their basic qualities of good neighbourliness and thus making them into rather good living environments for a new generation of owner-occupiers. But a generation later, at

Plate 2.1 Early 'industrial dwellings' in Bethnal Green, London. From the mid-nineteenth century onwards, these were a reaction by private philanthropic landlords to the slum problem of the Victorian city. They offered superior working-class accommodation and yet gave a return on capital.

the turn of the century, some of these areas – especially in northern cities – were being abandoned, as owners found newer and more desirable housing elsewhere.

Commonly, even with the more generous standards of street width which were required, by-law housing was built at net densities of about 50 houses to the acre (124/hectare). (The term 'net density', often used in this book, means density of housing or people on the actual housing area, including local streets; it does not include associated open space, public buildings or industry.) Given the large families of three or four children prevailing, this could mean densities of 250 people to the acre (620/hectare) or more. In London densities of over 400 to the acre (1,000/hectare) persisted in places as late as the Second World War. Today, with smaller households, these densities are a memory.

The phenomenon of urban spread

But by and large the period after 1870 marks a significant change in the development of British cities – and, as far as can be seen from international studies by the economist Colin Clark, in other countries' cities too. In fact, the trend is quite marked for London after the 1861 census. Up to that time, as we noted earlier, densities were actually rising within a radius of about 3 miles (4.8 kilometres) from the centre of British cities – the radius within which people could walk to their work within about an hour, there being no effective public or private transport of any kind for most of the population. If we look at a town like Preston (Plate 2.2), which had changed little in the 100 years or so between the time when most of the buildings were erected and the time of the photograph (about 1935), we should realize that most of the people living in these gardenless houses, without public parks, nevertheless could walk to open fields within about 20 minutes. (This was true in 1935 as in 1835.) And since the cotton mills – then the chief and almost the sole source of work for many – were scattered fairly evenly across the town, journeys to work on foot were quite extraordinarily short: an average mill hand could walk to and from work four times a day, coming home for a midday meal in rather less time than the average modern commuter spends on his or her outward morning journey. Even the biggest European city, London, grew relatively little in area as it doubled in population from 1 to 2 million people between 1801 and 1851.

But then, between about 1870 and 1914, virtually all British cities rapidly acquired a cheap and efficient public-transport system – first (in the 1870s and 1880s) in the form of horse trams and buses, then (about the turn of the century) of electric trams, and lastly (just before the First World War) in the form of motor buses. In very large cities like London there were also commuter trains. The early railways had neglected the possibilities of suburban traffic, even in London, but most of them awoke to the possibilities after 1860; and one, the Great Eastern serving north-east London, was compelled by Parliament to run cheap trains for workmen, allowing them to live in suburbs as distant as Edmonton and Leytonstone. London even had a steam-operated underground railway, the world's first, by 1863; its first electric tube railway opened in 1890 and its first electrified suburban lines in 1905–9.

The impact on urban growth was profound, as can clearly be seen in the series of maps for London at different dates (Figure 2.3). London in 1801, with 1 million people, was still a remarkably compact city, mainly contained within a radius of about 2 miles (3.2 kilometres) from the centre; and by 1851, with double the number, the radius had not increased to much more than 3 miles (4.8 kilometres), with higher densities in the inner areas. Then the city began to spread in all directions, but particularly to the south and north east – as seen in the map for 1880 and, even more clearly, for 1914. This last represents the apogee of what can fairly be called the early public-transport

Plate 2.2 Aerial photograph of Preston in the 1930s. This demonstrates the high density and closely built-up nature of the early industrial town. Though open space is lacking, the town is small, and open countryside is not far away (though not visible here); and, with factories scattered among houses, the journey to work is short. Today, many of the mills have been demolished, although those remaining are the subject of conversion and redevelopment into residential, commercial or community uses. Much of the housing in this picture has been replaced. Many of the inhabitants of the town doubtless travel farther to work, for many of the jobs are either on the town's periphery or farther afield in and around the Lancashire conurbation, accessed via motorway and rail services.

city. The steam trains gave fairly easy and rapid access to middle-class commuters (and, in east London, the working class too) at distances up to 15 miles (24 kilometres) from the centre. But they accelerated and decelerated poorly; stops tended to be widely spaced; and feeder services, in the form of horse buses or trams, were poorly developed, or slow. The result is a typically tentacular form of growth, with development taking the form of blobs (or beads on a string, to change the metaphor) around each station.

Between the two world wars the whole process of suburban growth and decentralization began to speed up; in doing so it changed its form. The forces behind the suburban movement during those years were partly economic, partly social, partly technological. Economic forces in the world outside – world depression between 1929 and about 1934, a general depression in the prices of primary products – meant that both labour for construction and building materials were cheap. Social changes, too, were produced by economic development: more and more workers were becoming white-collar employees in offices or shops or other non-factory occupations, enjoying regular salaries which allowed them to borrow money on credit, and regarding themselves as members of an enlarging middle class. In large numbers, these people began to aspire to buy a house of their own with the aid of a mortgage. Lastly and perhaps most fundamentally, further developments in transport technology extended the effective commuting range: electric

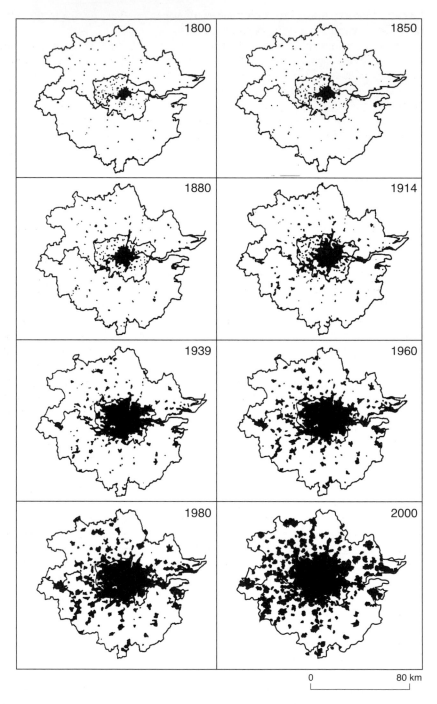

Figure 2.3 The growth of London, 1800–2000. Until 1850 London's extent was constrained by walking distances. Steam trains from 1850 to 1914, and electric trains, tubes and buses from 1914 to 1939, allowed suburban sprawl – but then the green belt stopped it.

trains in London, motor buses elsewhere, allowed the effective area of the city to extend up to four or five times the previous limits.

This was particularly well marked in London. In 1914 London had a population of about 6.5 million; by 1939, 8.5 million. Yet in that period, the capital's built-up area extended about three times. The underground railways before 1914 had barely extended beyond the existing developed area; but after 1918 they began to colonize new territory, extending quickly above ground onto previously undeveloped areas. The result was as predicted: a vast flood of speculative building, cheaply built for sale. Plate 2.3 shows the result around just one station: Edgware in Middlesex, some 12 miles (19 kilometres) from central London, in 1926 – two years after the line was opened – and 1948 – a quarter of a century later.

The precise impact of this sort of development upon the urban structure can be well seen by comparing the maps of London, in 1914 and 1939, respectively, in Figure 2.3. London in 1914, as we already noted, had the characteristically tentacular shape associated with the early public-transport city – the city of the steam train and the horse bus. By 1939 London had assumed a completely different shape: growth was much more even in any direction, producing a roughly circular city with a radius about 12 to 15 miles (19 to 24 kilometres) from the centre. The basic reason for this was a change in the technology of transportation. First, electric trains were more efficient carriers than the steam trains had been: accelerating and decelerating rapidly, they could serve more frequently spaced stations. Second, and even more importantly, the motor bus allowed a fairly rapid urban-transport service to penetrate in any direction from these stations, along existing roads, without the need for elaborate capital investment on the part of the operator; it therefore served as a highly efficient feeder service. These changes altered the pattern of accessibility within the urban area. The isochrones (lines of equal accessibility to the centre, in terms of time) were in 1914 very irregular; they fingered out a long way along the rail lines. By 1939 they had become more even and circular (or concentric) in form; and the development of the urban area followed accordingly. This form we can call typical of the later public-transport city; it was not at all a creation of the private car, since in London by 1939 only about one family in ten owned one.

The same process was repeated around the provincial cities too; it was merely on a smaller scale, and dependent on the tram or bus rather than the train. In some of the bigger cities – Manchester, Liverpool and Leeds – the local authorities themselves contributed to the process. They rehoused many thousands of slum dwellers and other people in need of public housing by developing new estates of single-family homes – generally at distances from 4 to 7 miles (7–11 kilometres) from the city centre, in the case of the biggest cities, and connected to it by rapid, frequent and cheap public transport. Like the private housing, this was cheaply built (and, unlike most of the private housing, it was aided by central-government subsidy as the result of a 1919 Housing Act). It was also of a standard never before reached in public housing: equipped with basic facilities like bathrooms, and with generous private garden space around. These authorities built fairly faithfully according to the recommendations of an influential official report, the Tudor Walters Report, which had been published at the end of the First World War in 1918; it had recommended development of single-family homes at about 12 per net residential acre (30/hectare), or about one quarter the density of the old by-law housing.

This also was the density of much of the private housing developed around London and other big cities; many private estates were built at even lower densities: 10 or 8 or even 6 houses to the acre (15–24/hectare). For the general feeling was that more spacious housing standards were a healthy reaction to the cramped terraces of the nineteenth-century industrial town; the bus and the electric train had liberated the manual workers in their rented council houses and the white-collar workers in their

Plate 2.3 Edgware, north-west London: (a) 1926; (b) 1948, showing the impact of the extension of the underground railway (station in centre of pictures) on suburban development. Typical are the uniform rows of semi-detached housing, built at 12 dwellings to the acre (30/hectare), with generous gardens. Better transportation allowed the city to spread.

mortgaged semi-detached house alike. And because the improved transportation made so much land potentially developable, the price of land was low. Indeed, it is clear from later research that land prices and the house prices, which are always so closely related, reached a low point in relation to income in the 1930s that has never been equalled before or since (Figure 2.4). It was actually easier for the average clerical or skilled manual worker to buy a house in the 1930s than it is in the more affluent Britain of 80 years later.

The reaction against sprawl

A minority of thinking people, however, were alarmed at the result. They included both town planners, who by then existed as a profession – the Town Planning Institute had been incorporated in 1914 – and rural conservationists. They were concerned at the fact that the development was uncontrolled by any sort of effective planning. Though Acts of Parliament had provided for local authorities to make town-planning schemes for their areas – in 1909, in 1925 and then, most decisively, in 1932 – basically these Acts gave them no power to stop development altogether where such development was not in the public interest; developers could build almost wherever they liked, provided they followed the general lines of the local town-planning scheme. And this, the planners and conservationists argued, had two bad effects.

First, it was using up rural land – the great majority of it agricultural land – at an unprecedented rate. By the mid-1930s, as subsequent research showed, some 60,000 acres (24,000 hectares) each year (out of 37 million acres (15 million hectares) in all) were being taken from agriculture in England and Wales for all forms of urban development. Because the development was completely uncontrolled, it was no respecter of the quality of agricultural land: the suburban spread of London, for instance, took much of the finest market-gardening land in all England, on the gravel terrace lands west of the capital (and ironically, later on, Heathrow Airport took much of the rest; see Plate 2.4). The result, critics argued, was a major loss in domestic food production – a loss Britain could ill afford in times of war. And in the late 1930s, with war threatening, this seemed an important argument.

Second, the critics argued that the effect on the townspeople was equally bad. Homes were being decentralized at greater and greater distances from the city centre, but jobs were not being decentralized nearly as rapidly. In London and in some of the bigger provincial cities, between the two world wars some factory industry was moving outwards

Figure 2.4 A house-agent's advertisement of the early 1930s. At this time house prices, aided by cheap labour and materials, were probably cheaper in relation to white-collar salaries than ever before or since. Commuting on the new electric lines round London was easy. There was a striking contrast with the poverty in the depressed industrial areas of the North.

Plate 2.4 The Great West Road, London, in 1951; ribbon development of the 1920s and 1930s alongside an interwar arterial road. This consumed most of the best agricultural land in southern England, and aided the movement in the 1930s for more effective controls on urban growth. It also compromised the original purpose of the road as a through road, so that by the mid-1960s a replacement motorway was needed.

to the suburbs in search of space: new factory estates were developed like Park Royal and the Lea Valley in London, Slough just outside it, Witton Park in Birmingham or Trafford Park in Manchester. But much industry remained in inner urban locations, and the growing volume of so-called tertiary industry – service occupations like work in offices and shops – seemed to be firmly locked in city centres. As a result, traffic congestion in the cities appeared to be growing; and journeys to work, it was assumed, must be getting longer all the time. As cities grew larger and larger, as their suburbs sprawled farther and farther, it was argued that they imposed an increasingly insufferable burden on their inhabitants. And as new arterial roads were built to relieve traffic congestion on the old radial arteries out of the city, so these in turn were lined by ribbon development of new housing, compromising their function and reducing their efficiency. Ribbon development was partially controlled by an Act of 1935, but the real answer to the problem – motorways for through traffic, with limited access, already being opened in Italy and Germany – was not introduced to Britain until the Special Roads Act of 1949.

Thus a small, but powerful and vocal, movement built up to limit urban growth through positive planning. Essentially, it represented a working coalition between people interested in town planning – some, but not all of them, professional planners – and rural preservationists, who had been instrumental in organizing the Council for the Preservation of Rural England (CPRE) in 1925. One strong figure spanned both camps and united them: Patrick Abercrombie, Professor of Planning at the University of London and founder of the CPRE. Though they were persuasive, they might not have been so effective if they had not been joined by a third group: the representatives of the depressed industrial areas of northern England, south Wales and central Scotland. We shall see in Chapter 4 how this happened. But meanwhile, we need to retrace our steps in time, to look at some of the most important ideas circulating among urban planners, and others interested in the subject, at this time.

Further reading

A standard textbook of modern economic and social history will provide indispensable background. Good examples include E.J. Hobsbawm, *The Pelican Economic History of Britain*, Vol. 3: *Industry and Empire* (Penguin, 1970); and P. Mathias, *The First Industrial Nation: An Economic History of Britain 1700–1914* (Methuen, 1969; paperback edition available). Roy Porter, *London: A Social History* (Hamish Hamilton, 1994) and Michael Hebbert, *London: More by Accident than Design* (Wiley, 1998) contain excellent expositions of London's interwar development.

These texts should be supplemented by W. Smith, *A Historical Introduction to the Economic Geography of Great Britain* (Bell, 1968), which emphasizes the geographical impact of economic change, and then by W.G. Hoskins, *The Making of the English Landscape* (Hodder & Stoughton, 1988), which discusses the impact on the landscape.

On the earlier history of town planning, see W. Ashworth, *The Genesis of Modern British Town Planning* (Routledge, 1954), and Leonardo Benevolo, *The Origins of Modern Town Planning* (Routledge, 1967). These should be supplemented by Colin and Rose Bell, *City Fathers* (Penguin, 1972), which gives an indispensable picture of early town planning experiments; and by Gordon E. Cherry, *Cities and Plans: The Shaping of Urban Britain in the Nineteenth and Twentieth Centuries* (Edward Arnold, 1988).

On the interwar development, useful sources are S.E. Rasmussen, *London the Unique City* (Cape, 1937), A.A. Jackson, *Semi-Detached London* (second edition, Wild Swan Publications, 1991), and M. Swenarton, *Homes Fit for Heroes* (Heinemann, 1981). Peter Hall, *Cities of Tomorrow* (Blackwell, 1988) is a comprehensive history of planning in this period. For London, see Roy Porter, *London: A Social History* (Hamish Hamilton, 1994) and Rob Imrie, Loretta Lees and Mike Raco, *Regenerating London: Governance, Sustainability and Community in a Global City* (Routledge, 2009).

3 The seers: pioneer thinkers in urban planning, from 1880 to 1945

During the whole of Chapter 2 we concentrated on the evolution of what can be called, broadly, the urban problem in Britain from the Industrial Revolution of the late eighteenth century to the outbreak of the Second World War. We looked at the facts of urban development and at the attempts – often faltering and not very effective ones – on the part of central and local administration to deal with some of the resulting problems. This was the world of practical men grappling with practical matters. But no less important, during this time, were the writings and the influence of thinkers about the urban problem. Often their writings and their lectures reached only a tiny minority of sympathetic people. To practical people of the time, much of what they asserted would seem utopian, even cranky. Yet in sum, and in retrospect, the influence of all of them has been literally incalculable; furthermore, it still continues.

This delay in the recognition and acceptance of their ideas is very important. Some of these ideas were more or less fully developed by the end of the nineteenth century, and a large part were known to the interested public by the end of the First World War. Yet with the exception of some small-scale experiments up to 1939, nearly all the influence on practical policy and design has come since 1945. One obvious peril in this is that no matter how topical and how appropriate these thinkers were in analysing the problems of their own age, their remedies might be at least partially outdated by the time they came to be taken seriously. We shall need to judge for ourselves how serious this has been.

It is useful to divide the thinkers into two groups: the Anglo-American group and the Continental European group. The basis of the distinction here is more than one of convenience. Basically the background of the two groups of thinkers has been quite different. We already saw in Chapter 2 that in England and Wales (Scotland in this respect has been rather more like the European Continent), cities began to spread out after about 1860: first the middle class and then (especially with the growth of public housing after the First World War) the working class began to move out of the congested inner rings of cities into single-family homes with individual gardens, built at densities of 10 or 12 houses to the acre (25–30/hectare). Exactly the same process occurred, from about the same time, in most American cities, though in some cases the process was delayed by the great wave of arrivals of national groups (such as Italians, Greeks, Russians, Poles and Jews) between 1880 and 1910; they crowded together in ethnic ghettos in the inner areas of cities like New York, Boston or Chicago, and took some time to join the general outward movement. Nevertheless, by the 1920s and 1930s there was a rapid growth of single-family housing around all American cities, served by public transport and then, increasingly, by the private car. This was a tradition which, by and large, writers and thinkers in both Britain and the United States accepted as the starting point.

On the Continent of Europe it was quite otherwise. As cities grew rapidly under the impact of industrialization and movement from the countryside, generally several decades later than the equivalent process in Britain (i.e. from about 1840 to 1900), they failed to spread out to anything like the same extent. As public-transport services developed, generally in the form of horse- and then electric-tram systems, some of the middle class, and virtually all the working class, continued to live at extraordinarily high densities virtually within walking distance of their work. The typical Continental city consisted then, and still consists today, of high apartment blocks – four, five or six storeys high – built continuously along the streets, and thus enclosing a big internal space within the street block. In middle-class areas this might be a pleasant communal green space; in other areas it was invariably built over in the desperate attempt to crowd as many people as possible in. The result by 1900 was the creation of large slum areas in most big European cities, but of a form quite different from the English slums. In England even poor people lived in small – generally two-storey – houses of their own, either rented or bought. In Europe they lived in small apartments, and the densities – both in terms of persons per room, and dwellings or persons per net residential acre – were much higher than in typical English slum areas. (Scotland, curiously, developed in the European way: Glasgow, for instance, is a city of tenements, not houses, and standards of crowding have always been much worse there than in big English cities.) Naturally, when Continental Europeans began to think about urban planning, they tended to accept as a starting point this apparent preference for high-density apartment-living within the city.

The Anglo-American tradition

Howard

The first, and without doubt the most influential, of all the thinkers in the Anglo-American group is Ebenezer Howard (1850–1928). His book *Garden Cities of To-morrow* (first published in 1898 under the title *To-morrow*, and republished under its better-known title in 1902) is one of the most important books in the history of urban planning. Reprinted several times and still readily available as a paperback, it remains astonishingly topical and relevant to many modern urban problems. From it stems the whole of the so-called garden city (or in modern parlance, new town) movement which has been so influential in British urban planning theory and practice.

To understand its significance it is necessary to look at its historical background. Howard was not a professional planner – his career, if he can be said to have one, was as a shorthand writer in the law courts – but a private individual who liked to speculate, write and organize. As a young man he travelled, spending a number of years in the United States during its period of rapid urban growth before returning to England to write his book. At that time several pioneer industrialists with philanthropic leanings had already started new communities in association with large new factories which they had built in open countryside. (Their motives, perhaps, were not entirely philanthropic: they built their factories cheaply on rural land; it was necessary to house the labour force outside the city in consequence, and they got a modest return in rents for their investment.) The earliest of these experiments, Robert Owen's celebrated experimental settlement at New Lanark in Scotland (*c.* 1800–10) and Titus Salt's town built round his textile mill at Saltaire, near Bradford (1853–63), actually date from the early years of the Industrial Revolution (see Plate 3.1a, b). But the best known and the most important date from the late nineteenth century, when the growing scale of industry was tending to throw up a few very powerful industrialists who saw the advantages of decentralizing their plants far from the existing urban congestion. Bournville, outside Birmingham (1879–95), built by the chocolate manufacturer George Cadbury, and Port Sunlight, on the Mersey near Birkenhead (1888, built by the chemical magnate William Hesketh Lever), are the best-known examples in Britain (Plate 3.1c, d). In Germany the engineering and armaments firm, Krupp, built a number of such settlements outside their works at Essen in the Ruhr district, of which the best preserved, Margarethenhöhe (1906),

(a)

(b)

Plate 3.1 (a) New Lanark (Robert Owen, *c.* 1800–1810); (b) Saltaire (Titus Salt, 1853–63); (c) Bournville (George Cadbury, 1879–95); (d) Port Sunlight (William Lever, 1888). These four pioneer new towns were established by philanthropic industrialists around their works in open countryside.

(c)

(d)

Plate 3.1 continued

closely resembles Bournville and Port Sunlight. Similarly, in the United States the railroad engineer George Mortimer Pullman (who invented Pullman cars) built a model town named after himself, outside Chicago, from 1880 onwards.

These towns all contain the germ of the idea which Howard was to propagate: in all of them industry was decentralized deliberately from the city, or at least from its inner sections, and a new town was built around the decentralized plant, thus combining working and living in a healthy environment. They are, in a sense, the first garden cities, and many of them are still functional and highly pleasant towns today. But Howard generalized the idea from a simple company town, the work of one industrialist, into a general planned movement of people and industry away from the crowded nineteenth-century city. Here he drew on previous writings: on Edward Gibbon Wakefield, who had advocated the planned movement of population even before 1850, and James Silk Buckingham, who had developed the idea of a model city. But perhaps the strongest intellectual influence on Howard's thinking was that of the great Victorian economist Alfred Marshall; he, if anyone, invented the idea of the new town as an answer to the problems of the city, and he gave it an economic justification which only later came to be fully understood. Marshall argued, as early as 1884, that much industry was even then footloose, and would locate anywhere if labour was available; he also recognized that the community would eventually have to pay the social costs of poor health and poor housing, and that these were higher in large cities (as they then existed) than they would be in new model communities.

Howard, however, developed the idea, generalized it and above all turned it into an eminently practical call for action. And of all visionary writers on planning, Howard is the least utopian, in the sense of impractical; his book is packed with detail, especially financial detail, of how the new garden cities were to be built. But first of all Howard had to provide a justification of the case for new towns (or garden cities) that could be readily understood by practical men without much knowledge of economics. He did so in the famous diagram the Three Magnets (Figure 3.1), which in fact is an extremely compressed and brilliant statement of planning objectives. (It is an interesting exercise to try to write out the diagram in suitably jargon-ridden, abstract modern language as a statement of objectives; to say the same thing less clearly takes many pages, whereas Howard got it all in one simple diagram.) Basically, Howard was saying here that both existing cities and the existing countryside had an indissoluble mixture of advantages and disadvantages. The advantages of the city were the opportunities it offered in the form of accessibility to jobs and to urban services of all kinds; the disadvantages could all be summed up in the poor resulting natural environment. Conversely, the countryside offered an excellent environment but virtually no opportunities of any sort.

It is important here to remember the date of Howard's book. In the 1890s material conditions in British cities were better than they had been in the 1840s. Average incomes for many workers were significantly higher; medical standards had improved; and the new housing by-laws were beginning to have effect. Nevertheless, by modern standards they were still appalling. The 1891 census showed that at least 11 per cent of the population, over 3 million people, were living at densities of over 2 persons per room; and this was certainly an underestimate. Even in the 1880s the Registrar General's records showed that the expectation of life in a city like Manchester was only 29 years at birth on average – only 5 years more than 40 years previously. In the late 1880s and early 1890s the shipowner Charles Booth conducted the first modern social investigation, based on strict statistical recording. Aided by the young Beatrice Webb, he produced a study which is still a classic: it showed that on a strict and minimal standard no less than one quarter of the population of inner East London was living below the poverty line. But on the other hand there was equal distress in the countryside: these were the

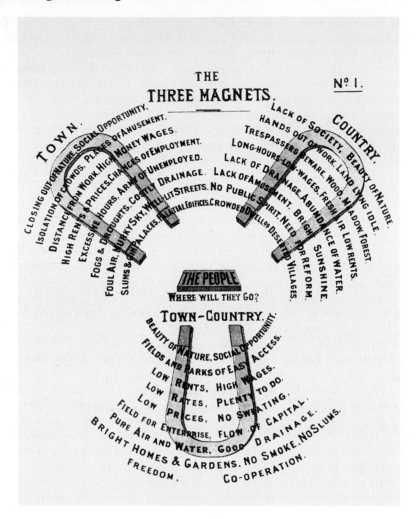

Figure 3.1 Ebenezer Howard's Three Magnets. The celebrated diagram from *Garden Cities of To-morrow* (first published in 1898) setting out the advantages and disadvantages of town and country life. A hybrid form of the future, the planned Town–Country or Garden City, combined the advantages of both with none of the disadvantages – so Howard argued.

years of deep agrarian depression, brought about by the mass importation of cheap foreign meat and wheat against which the British farmer was given no protection. The population map of Britain in the 1890s shows losses almost everywhere except for the limited areas of the cities and the industrial districts. Though the towns were beginning to spawn suburbs, there was virtually none of the twentieth-century phenomenon whereby urban workers could afford to live in the countryside; that had to wait for the motorcar. And when Howard's book was published, it was precisely two years since Parliament had removed the requirement that a car must be preceded along the highway by a man with a red flag. He could, perhaps, hardly be expected to foresee the consequences of liberating the car.

Against this background Howard argued that a new type of settlement – Town–Country, or Garden City – could uniquely combine all the advantages of the town by way of accessibility, and all the advantages of the country by way of environment, without any of the disadvantages of either. This could be achieved by planned decentralization of workers and their places of employment, thus transferring the advantages or urban agglomeration *en bloc* to the new settlement. (In modern economic jargon, this would be called 'internalizing the externalities'.) The new town so created would be deliberately outside normal commuter range of the old city. It would be fairly small – Howard suggested 30,000 people – and it would be surrounded by a large green belt, easily accessible to everyone. Howard advised that when the town was established 6,000 acres (2,400 hectares) should be purchased: of this no less than 5,000 acres (2,000 hectares) would be left as green belt, the town itself occupying the remainder.

Two important points about Howard's idea especially need stressing, because they have been so widely misunderstood. The first is that, contrary to the usual impression, Howard was advocating quite a high residential density for his new towns: about 15 houses per acre, which in terms of prevailing family size at the time meant about 80–90 people per acre (200–220/hectare). (Today it would mean 40–50.) The second is that he did not advocate small, isolated new towns. His notion was that when any town reached a certain size, it should stop growing and the excess should be accommodated in another town close by. Thus the settlement would grow by cellular addition into a complex multicentred agglomeration of towns, set against a green background of open country. (And even this was to be fairly densely populated by space-consuming urban activities like public institutions: Howard allowed for one person to every 4 acres (2 hectares) there.) Howard called this polycentric settlement the 'Social City'. The diagram in the first edition of his book showed it as having a population of 250,000 – the original target population of the modern 'giant' new town of Milton Keynes in England – but Howard himself stressed that the Social City could grow without limit. This point has never been well understood, because the second edition of Howard's book, and all subsequent editions, have omitted the diagram; it is reproduced here in Figure 3.2.

Howard, as we have noticed, was very specific about how his new communities could be built. Private enterprise could do it, he stressed, if money could be borrowed for the purpose: land could be bought cheaply in the open countryside for the project, and the subsequent increase in land values would allow the new town company to repay the money in time and even make a profit to be ploughed back into further improvement, or into the creation of further units of the Social City. In fact, Howard was actually instrumental in getting two garden cities started – Letchworth in northern Hertfordshire (1903) and Welwyn Garden City a few miles to the south (1920). Both were built very much on the lines he advocated, with wide green belts around (Figure 3.3 and Plate 3.2). But both suffered financial troubles, and the vision of private-enterprise new towns on a large scale was never realized. Furthermore, despite insistent and effective propaganda from the Town and Country Planning Association, which he founded, governments after the First World War failed to respond to the call for public new towns.

Unwin and Parker

Between 1900 and 1940 many of Howard's ideas were developed by his faithful followers. Among the most prolific and brilliant of the writers was Sir Frederic Osborn (1885–1978), who lived to see over a score of new towns built in England after the Second World War. In terms of physical realization the opportunities were clearly more limited. The two architects who designed the first garden city, Letchworth, Raymond Unwin (1863–1940) and his assistant Barry Parker (1867–1947), later went on to build Hampstead Garden Suburb at Golders Green in north-west London (1905–9). As its

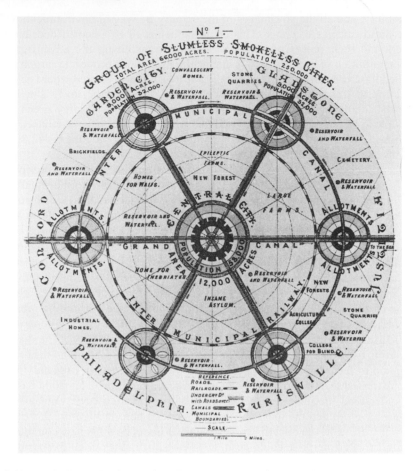

Figure 3.2 Ebenezer Howard's Social City. The lost diagram from the first edition of Howard's book, demonstrating the full conception of garden cities (or new towns) grouped in planned urban agglomerations of a quarter of a million people or more.

name indicates, this was not a garden city but a dormitory suburb owing its existence to the new underground line opened in the year 1907; and it was condemned by many garden city supporters on that ground. But it was an interesting experiment in the creation of a socially mixed community, with every type of house from the big mansion to the small cottage; and in its creation of a range of houses which are all skilfully designed, all varied yet all quietly compatible, it is one of the triumphs of twentieth-century British design.

Later, Parker went on to a more ambitious enterprise: the design of a new community for 100,000 people to be built by the city of Manchester at Wythenshawe, south of the city (1930). Wythenshawe in fact deserves to be called the third garden city (or new town) actually started in Britain before the Second World War. It has all the essential features of the design of Letchworth or Welwyn: the surrounding green belt, the mixture of industrial and residential areas, and the emphasis on single-family housing of good design. (The family resemblance between Hampstead Garden Suburb and Wythenshawe is more than coincidental.) It did, however, compromise on the principle of self-containment: because most of its public transport was provided for them to commute

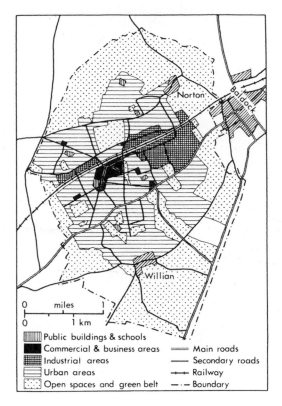

Figure 3.3 Plan of Letchworth Garden City, 1903. This was the first garden city, built in northern Hertfordshire with private capital under Howard's general direction. The architects were Raymond Unwin and Barry Parker.

back. But the intention – never completely realized in practice – was to provide a wide range of jobs in the community itself.

Together, Unwin and Parker developed some important modifications of the original Ebenezer Howard idea. In a very influential pamphlet published in 1912, *Nothing Gained by Overcrowding!*, Unwin argued that housing should be developed at lower densities than were then common. The need for public open space, he pointed out, was related to the numbers of people, so that the saving in land from higher urban densities was largely illusory. He recommended a net density in new residential areas of about 12 houses to the acre (30/hectare) – or, in terms of the average family size of the time, about 50–60 people to the acre (124–150/hectare). This standard was accepted in the important official Tudor Walters Report of 1918, as we saw, and became usual in most public housing schemes of the 1920s and 1930s: Wythenshawe, like many other major schemes by city housing departments, was built at about this density.

Both Unwin and Parker consistently argued for the Howard principle of generous green belts around the new communities. In Unwin's graphic term, used in the regional plan he produced for the London area in the late 1920s, they would be cities against a background of open space – not cities surrounded by green belts, in the conventional use of the term. But Parker developed the idea still further. Visiting the United States in the 1920s, he was impressed by the early experiments in building parkways, i.e. scenic roads running through landscaped open country. Parker argued that the 'background of open space' between cities should be occupied by these parkways, giving easy inter-connection between them; this in fact was an adaptation to the motor age of Howard's original idea (shown in Figure 3.4a) of an inter-urban railway. Parker's conception of the parkway is shown in Figure 3.4b; he actually managed to half-build one in the middle

Plate 3.2 Letchworth from the air, showing the general physiognomy of the town. The large open green space in the centre of the town faithfully followed the original schematic plan in Howard's book. Industry is aligned along the railway, which existed before work on the town was started.

of Wythenshawe (see Figure 3.4c), and it was later completed as the M56 North Cheshire motorway, though not as he would have intended it.

Lastly, at Wythenshawe, Parker employed yet another notion he had picked up in the United States, which was in fact a logical development of Howard's own ideas: the idea of dividing the town into clearly articulated neighbourhood units. The ground plan of

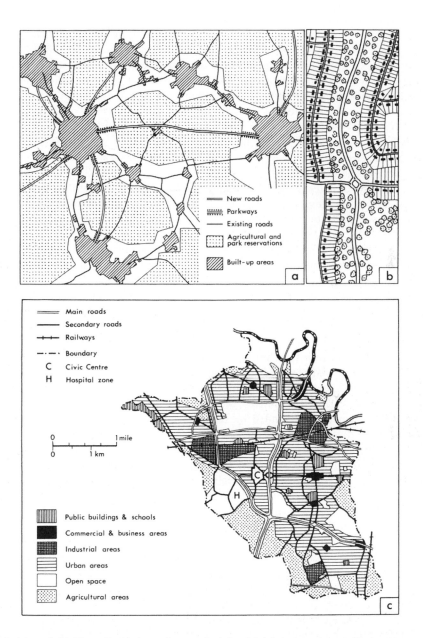

Figure 3.4 (a) and (b) Barry Parker's parkway principle; (c) its expression at Wythenshawe, 1930. Parker, Unwin's assistant for Letchworth Garden City, later developed the idea of the parkway: a landscaped road running through wedges of green space between strips of urban development. He tried to apply the idea in the centre of his satellite town of Wythenshawe, for the city of Manchester, but it was not completed according to his original conception.

Wythenshawe as actually completed shows the influence of this idea. To see its origins, we now need to follow across the Atlantic the Anglo-American tradition of thought.

Perry, Stein and Tripp

In Howard's original theoretical diagram of his Garden City, published in 1898, he divided the town up into 'wards' of about 5,000 people, each of which would contain local shops, schools and other services. This, in embryo, is the origin of the neighbourhood-unit idea, which in essence is merely pragmatic: certain services, which are provided every day for groups of the population who cannot or do not wish to travel very far (housewives and young children), should be provided at an accessible central place for a fairly small local community, within walking distance of all homes in that community. Depending on the residential density, the idea of convenient walking distance will dictate a limit of a few thousand people for each of these units. It makes psychological sense to give such a unit a clear identity for the people who live in it, by arranging the houses and streets so that they focus on the central services and providing some obvious boundary to the outside.

In the United States, however, the idea was taken much further during the preparation of the New York Regional Plan in the 1920s. (This great multi-volume plan, prepared wholly by a voluntary organization, is one of the milestones of twentieth-century planning.) One contributor to this plan, Clarence Perry (1872–1944), there developed the idea of the neighbourhood unit, not merely as a pragmatic device, but as a deliberate piece of social engineering which would help people achieve a sense of identity with the community and with the place. (He based it on a model garden suburb, Forest Hills Gardens in New York City, which he had helped to plan in 1912.) For this there was no empirical justification, and not much has emerged since; though some important work done for the Royal Commission on Local Government in England, and published in 1967, did suggest that most people's primary sense of identification was to a very small local area. But just in terms of physical planning Perry's work did give firmness to the neighbourhood idea. He suggested that it should consist of the catchment area of a primary school, extending about half or three-quarters of a mile in any direction, and containing about 1,000 families – or about 5,000 people, in terms of average family size then. It would be bounded by main traffic roads, which children should not be expected to cross (Figure 3.5).

The essential idea of the neighbourhood unit, as developed by Perry – though not some of his details, such as putting shops at the corners of the units at the junctions of traffic roads – was enthusiastically taken up by British planners in new towns and in some cities after the Second World War; its influence is everywhere to be seen. Since the early 1960s, however, it has come in for increasing criticism. An influential paper published in 1963 by Christopher Alexander, a young English émigré to the United States, called 'A City Is Not a Tree', suggested that sociologically the whole idea was false: different people had varied needs for local services, and the principle of choice was paramount. In his view cities that had grown naturally demonstrated a more complex settlement structure, with overlapping fields for shops and schools; planners should aim to reproduce this variety and freedom of choice. The master plan for the new town of Milton Keynes, published in 1969, was one of the first to reflect these ideas.

Meanwhile a close associate of Perry – Clarence Stein (1882–1975), an architect–planner working in the New York region – had taken the neighbourhood concept further. Stein was one of the first physical planners, apart from Parker in England and Le Corbusier in France, to face fully the implications of the age of mass ownership of the private car. He grasped the principle that in local residential areas the need above all

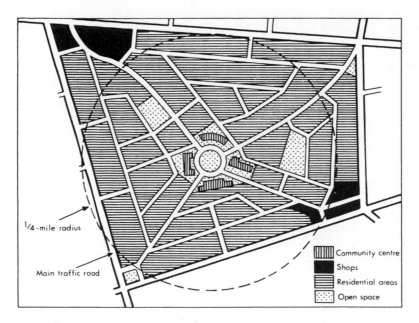

Figure 3.5 The 'neighbourhood unit' principle. First developed by the American architect–planner Clarence Perry in the celebrated New York Regional Plan of the 1920s, this principle was based on the natural catchment area of community facilities such as the primary schools and local shops. It was copied by Parker at Wythenshawe and then widely in British plans after the Second World War.

was to segregate the pedestrian routes used for local journeys – especially by housewives and children – from the routes used by car traffic. In a new town development at Radburn, northern New Jersey (1933), which was started but never completed, he applied these ideas by developing a separate system of pedestrian ways, reached from the back doors of the houses, which pass through communal open space areas between the houses, and thence cross under the vehicle streets. The vehicle streets, in turn, are designed according to a hierarchical principle, with main primary routes giving access to local distributors and then in turn to local access roads designed on the cul-de-sac (dead-end) principle, serving small groups of houses (Plate 3.3). The Radburn Layout, as it came to be known, was applied by Stein in one or two other developments in the United States in the 1930s, but was adopted in Britain only after the Second World War; in fact, most of the examples date from the late 1950s or later. Few of them have the charm and ease of the design in the surviving section of Radburn itself.

The transfer of Stein's ideas back across the Atlantic to Britain, there to be combined with Perry's neighbourhood idea, came via a curious route. In 1942 an imaginative Assistant Commissioner of Police (Traffic) at London's Scotland Yard called H. Alker Tripp (1883–1954) published a slim book called *Town Planning and Traffic*. Though there is no direct evidence that Tripp knew of Perry's or Stein's work, it seems possible that he had read of it. The most novel suggestion in the book was the idea that after the war British cities should be reconstructed on the basis of precincts. Instead of main city streets which served mixed functions and which had many points of access to local streets, thus giving rise to congestion and accidents, Tripp argued for a hierarchy of roads in which main arterial or sub-arterial roads were sharply segregated from the local streets, with only occasional access, and also were free of direct frontage development. These high-capacity, free-flow highways would define large blocks of the city, each of

Plate 3.3 Two views of Radburn, New Jersey, USA, a town designed in the early 1930s by the American architect Clarence Stein. This was the first recorded case of planned segregation of pedestrians from vehicle traffic, and gave its name to the Radburn Layout, widely used in British plans from the late 1950s onwards.

which would have its own shops and local services. Tripp illustrated the idea graphically in his book by applying it to an outworn (and heavily bombed) section of London's East End (Figure 3.6).

Tripp's book came at an opportune time. For at that point Patrick Abercrombie – the most notable professional planner in Britain of that age – was working with the chief architect of the London County Council, J.H. Forshaw, and his brilliant assistant Wesley Dougill, on a postwar reconstruction plan for London. In an important section of the plan (which was published in 1943) Abercrombie and Forshaw called for the widespread application of the precinctual principle to London, and illustrated its application to two critical areas where

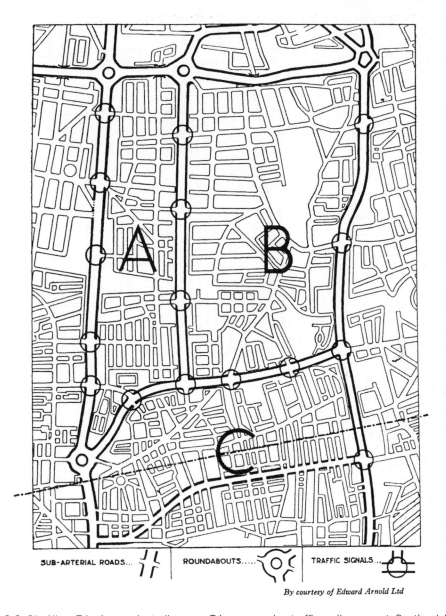

SUB-ARTERIAL ROADS...   ROUNDABOUTS.....   TRAFFIC SIGNALS...

By courtesy of Edward Arnold Ltd

Figure 3.6 Sir Alker Tripp's precinct diagram. Tripp, a senior traffic policeman at Scotland Yard, applied his ideas to a part of London's East End. Main roads would be largely sealed off from local side access, to give better traffic flow and safety; residential areas would be protected from heavy traffic. This arrangement would later be questioned by practitioners of the 'new urbanism'.

traffic threatened the urban life and fabric: the zone around Westminster Abbey, and the university quarter of western Bloomsbury (Figure 3.7). Ironically, in neither area were the ideas ever applied; indeed, at the end of the 1950s a new one-way traffic scheme actually routed through-traffic through the heart of Abercrombie's Bloomsbury precinct. But elsewhere – most notably, in the postwar reconstruction of the centre of Coventry, one of Britain's most heavily bombed cities – the idea was employed to good effect.

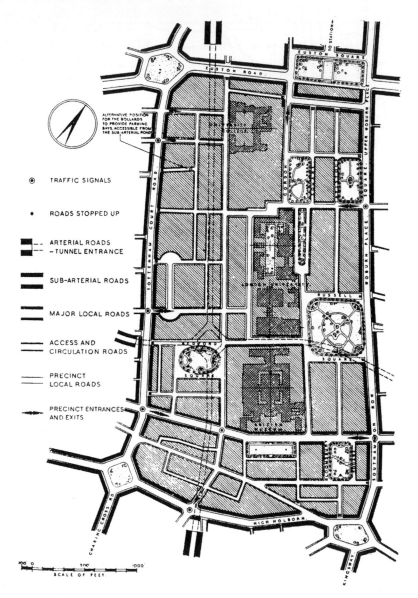

Figure 3.7 Abercrombie's Bloomsbury Precinct, from the County of London Plan 1943. Patrick Abercrombie and J.H. Forshaw applied Tripp's principles to the area around the British Museum and the University of London. Some of the necessary works were completed, but the idea was not fully realized; some 70 years later traffic is actually channelled through the precinct in a one-way system.

Geddes and Abercrombie

Abercrombie's most notable contributions to Anglo-American planning theory and practice, however, were made in extending city planning to a wider scale: the scale which embraced the city and the whole region around it in a single planning exercise. To understand this tradition and the way Abercrombie fits into it, it is necessary to go back some way and consider how the Ebenezer Howard tradition developed in another, slightly different direction.

From 1883 to 1919 a visionary Scots biologist, Patrick Geddes (1854–1932), taught at the University of Dundee, but did much of his most important work at his famous outlook tower in Edinburgh. Geddes's extraordinary mind soon took him away from conventional biology, into the area we should now recognize as human ecology: the relationship between humankind and its environment. In turn he was led to a systematic study of the forces that were shaping growth and change in modern cities, which culminated in his masterpiece *Cities in Evolution* (published in 1915, but mostly written about 1910). To understand the nature of Geddes's achievement, as with Howard's, it is necessary to place the book in the context of its time. Human geography, which had developed so finely in France during the first decade of the twentieth century in the hands of such practitioners as Vidal de la Blache and Albert Demangeon, was an almost unknown study in Britain; only one man, H.J. Mackinder, kept the subject alive, at the University of Oxford. But Geddes was fully acquainted with this tradition of study, and with the associated work of the French sociologist P.G.F. le Play. Both stressed the intimate and subtle relationships which existed between human settlement and the land, through the nature of the local economy; in le Play's famous triad, the relationship Place–Work–Folk was the fundamental study of people living in and on their land.

Geddes's contribution to planning was to base it firmly on the study of reality: the close analysis of settlement patterns and local economic environment. This led him to go right outside the conventional limits of the town, and to stress the natural region – a favourite unit of analysis of the French geographers – as the basic framework for planning. Today, when so many students are trained in the basic principles of human geography at school, all this seems very obvious and familiar. But, published at a time when planning for most practitioners was the study of civic design at a quite local level – a sort of applied architecture – it was quite revolutionary. Howard had already anticipated the change of scale; his analysis of the problem, and its solution, was a regional one. Geddes's contribution was to put the flesh of reality on the bare bones of the regional idea: at last, human geography was to provide the basis of planning. From this came Geddes's working method, which became part of the standard sequence of planning: survey of the region, its characteristics and trends, followed by analysis of the survey, followed only then by the actual plan. Geddes, more than anyone, gave planning a logical structure.

But his contribution did not end there. His analysis of cities in evolution led him to what was then a novel conclusion. Suburban decentralization, we saw in Chapter 2, was already by then causing cities to spread more widely. But in addition certain basic locational factors – the pull of coalfields in the early nineteenth century, the natural nodality conferred on certain regions by the way railways, roads and canals followed natural routes, the economies of scale and agglomeration in industry – had already caused a marked concentration of urban development in certain regions, such as the West Midlands, Lancashire and Central Scotland in Britain, or the Ruhr Coalfield in Germany. Geddes demonstrated that in these regions suburban growth was causing a tendency for the towns to coalesce into giant urban agglomerations or conurbations – the first time the latter word was used in the English language.

The conclusion Geddes drew was a logical one: if this was happening and would continue to happen, under the pressure of economic and social forces, town planning must be subsumed under town and country planning, or planning of whole urban regions encompassing a number of towns and their surrounding spheres of influence. Howard and his supporters had already drawn the same conclusion; and between the two world wars, aided powerfully by the persuasive writing of Geddes's American follower Lewis

Mumford – whose 1938 text *The Culture of Cities* became almost the bible of the regional planning movement – the idea gained a great deal of credence among thinking planners and administrators. Unwin was commissioned to prepare an advisory plan for London and its region, though funds ran out in the depression of 1931–3, before it could properly be completed. Already, here Unwin was applying the ideas of Howard to a planned schemer for large-scale decentralization of people and jobs from London to satellite towns in the surrounding Home Counties.

In this plan can be seen the germ of the Greater London Plan of 1944, which Patrick Abercrombie (1879–1957) prepared at the direct request of the British government. (It is significant that in this case, as elsewhere, extraordinary arrangements had to be made to prepare even an advisory plan for a whole urban region; the existing machinery of local government was quite inadequate in scale for the purpose.) But Abercrombie's great achievement was to weld this complex of ideas, from Howard through Geddes to Unwin, and turn them into a graphic blueprint for the future development of a great region – a region centred on the metropolis but extending for 30 miles (50 kilometres) around it in every direction, and encompassing over 10 million people. The broad aim of the plan was essentially Howard's: it was the planned decentralization of hundreds of thousands of people from an overcrowded giant city and their re-establishment in a great series of new planned communities, which from the beginning would be self-contained towns for living and working. The method was essentially Geddes's survey of the area as it was, including the historical trends which could be observed, followed by systematic analysis of the problem, followed by production of the plan. But the great sweep of the study, its characteristic assurance and its quality of almost cartoon-like clarity were essentially Abercrombie's own. However, the *Greater London Plan* essentially belongs to the story of the development of British planning at the time of the Second World War, which we treat in Chapter 4. We shall save a full discussion of it until then.

Wright

For the last important figure in the Anglo-American tradition we have to return across the Atlantic. Frank Lloyd Wright (1867–1959) does not fit readily into the line of development outlined in the previous pages, or into any line at all. It is fitting to put him last in this series, because his ideas about urban planning are so fundamentally at variance with those of the Continental school. Above all, they stand at the opposite extreme from those of Le Corbusier – the only other master of modern architecture whose ideas on planning are significant.

As with Corbusier, Wright's best-known monuments are his individual buildings, several of which are milestones of the modern movement; his ideas for planning on a wider scale never got further than paper. But unlike Corbusier, Wright's ideas were never taken up enthusiastically by a large following, either in Europe or in his native United States. They have, however, continued to exercise an important hold on a few influential thinkers in American planning practice during the 1950s and 1960s, especially in California. This is just, because in the same way that Corbusier's ideas are quintessentially European, so Wright's are typically American.

Wright based his thinking on a social premise: that it was desirable to preserve the sort of independent rural life of the homesteaders he knew in Wisconsin around the 1890s. To this he added the realization, based on the early spread of the motorcar among the farmers of North America, that mass car use would allow cities to spread widely into the countryside. With the car and with cheap electric current everywhere, Wright argued, the old need for activities to concentrate in cities had ended: dispersion, not only of homes but also of jobs, would be the future role. He proposed to accept this and to encourage it by developing a completely dispersed – though planned – low-density urban spread, which he called 'Broadacre City'. Here, each home would be surrounded by an acre of land, enough to grow crops on; the homes would be connected by super-highways, giving easy and fast travel by car in any direction. Along these highways he proposed a planned roadside civilization, in which the petrol ('gas') station would grow naturally into the emporium for a whole area; thus he anticipated the out-of-town shopping centre some 20 years before it actually arrived in North America (Figure 3.8). In fact, Wright's description of Broadacre City proved to be an uncannily accurate picture of the typical settlement form of North America after the Second World War – except that today the big half-acre or 1-acre (0.2–0.4-hecatre) lots grow very little food to support their families. The form developed without the underlying social basis that Wright so devoutly hoped for.

The European tradition

Planning as a tradition in Europe goes back to the Ancient Greeks. In the nineteenth century it produced such celebrated designs as the reconstruction of Paris under Georges-Eugène Haussmann (1890–91), which imposed a new pattern of broad boulevards and great parks on the previous labyrinthine street pattern. But since our task here is to understand how new ideas transformed town planning into city-regional planning, we again look at the visionaries.

Soria y Mata

Like many other thinkers considered here, the first representative of the opposing European tradition, the Spanish engineer Arturo Soria y Mata (1844–1920), owes his place in history to the importance of one basic idea. In 1882 he proposed to develop a linear city (*La Ciudad lineal*), to be developed along an axis of high-speed, high-intensity transportation from an existing city (Figure 3.9). His argument was that under the influence of new forms of mass transportation, cities were tending to assume such a linear form as they grew – an argument which, as we already saw in Chapter 2, had some justification at that time. Soria y Mata's ideas were ambitious if nothing else: he proposed that his linear city might run across Europe from Cadiz in Spain to St Petersburg in Russia, a total distance of 1,800 miles (2,900 kilometres). In fact, he succeeded only in building a few kilometres just outside Madrid; these still survive, though they are difficult to pick out on the map or on the ground, because they have been swallowed up in the amorphous growth of the modern city. In some ways the form seems archaic today: a main road runs straight through the linear centre of the city, carrying a tramway (since scrapped), with rather geometrical housing blocks on either side. And there seems no doubt from experience that such a form is difficult and costly to build; furthermore, even though commuter journeys may be fast, they are certainly likely to be long.

Nevertheless, the idea has always enjoyed some popularity among planners on the grounds that it has some good qualities. It does correspond to the need to exploit costly investments in new lines of rapid communication, whether these are nineteenth-century railways or twentieth-century motorways. (In both these cases, though, the settlement form that is most likely to arise is not linear, but rather a series of blobs round the stations or interchanges on the high-speed route.) And it does give easy access to nearby open countryside. Furthermore, it can respond automatically to the need for further growth, by simple addition at the far end; it does not need to operate through restrictive green belts, as Ebenezer Howard's finite garden has to. So it is not surprising that the

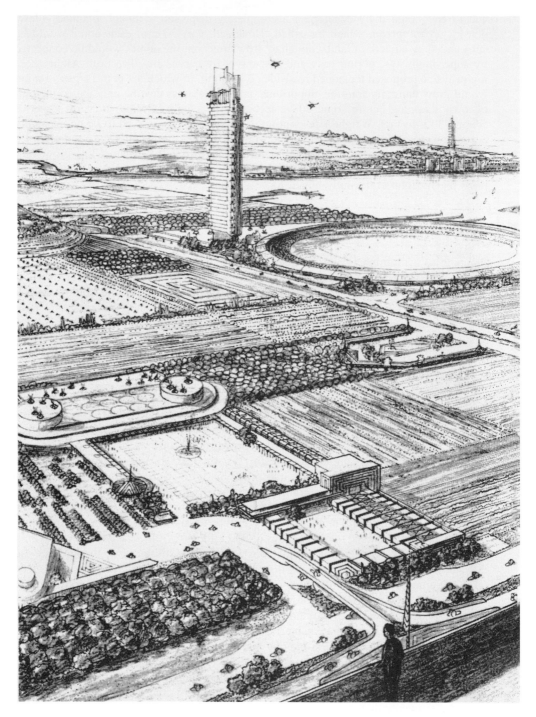

Figure 3.8 Broadacre City – the planning concept of the celebrated American architect and planner Frank Lloyd Wright in the 1930s. Single-family homes, each surrounded by an acre of land, allow each family to grow food for its own consumption. Transportation is by car, and the petrol ('gas') station becomes the focus of shopping and services. The concept is in sharp contrast to the ideas of European planners like Corubiser.

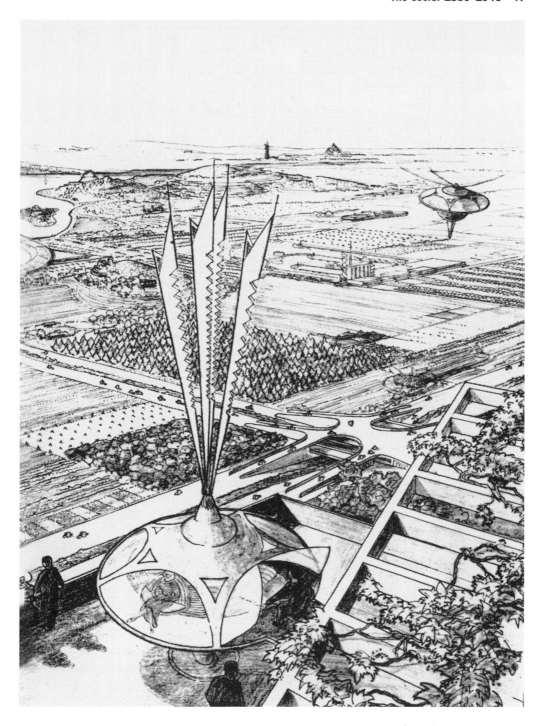

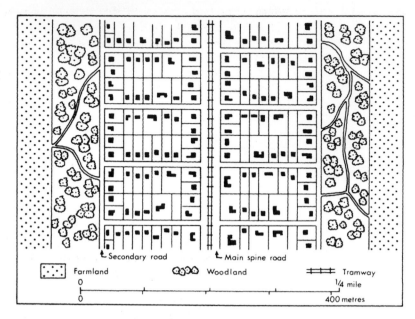

Figure 3.9 The Linear City, 1882. This concept of the Spanish architect Arturo Soria y Mata, based on a central rapid transit system, was actually begun outside Madrid but was swallowed up in the general development of the city. It has been influenced in many twentieth-century urban plans.

form has often appeared in regional plans as the most obvious alternative to the Howard–Abercrombie tradition. Parker's parkway is one example: the well-known MARS plan for London (1943), produced by a group of architects, used it; variants of it, in the postwar period, have appeared in plans for Copenhagen (1948), Washington (1961), Paris (1965) and Stockholm (1966). Some of these plans are discussed in more detail in Chapter 7. But one point can be made here: both in Washington and in Paris it proved extremely difficult to preserve the plan in the face of private attempts to build in the spaces left between the fingers or axes of urban growth. The claim that the linear city is a natural form, therefore, does not seem justified.

Garnier and May

The garden city was soon exported across the Channel. But curiously, in France its best-known expression seems to have occurred spontaneously, at about the same time that Howard was writing, without any mutual interaction. Tony Garnier (1869–1948), an architect working mainly in the city of Lyon, produced in 1898 – the same year precisely that Howard's was published – a design for an industrial city (*Cité industrielle*) which, like Howard's garden city, was to be a self-contained new settlement with its own industries and housing close by. The actual site Garnier chose to illustrate his scheme was just outside Lyon, and on it Garnier placed a rather strange elongated town, developed on a linear grid. The site plan, then, was unoriginal; but the detailing, with single-family houses in their own gardens, was remarkable for the France of that time. Furthermore, and perhaps the most striking feature, Garnier's houses made full use of new techniques of concrete construction – anticipating, in some ways, the designs of Le Corbusier over 20 years later.

The *Cité industrielle* was never built, and though garden cities were built around Paris during the 1920s and 1930s, most of them diverged in practice from their English models. They contained a high proportion of apartment blocks, and were remarkable mainly for their freer use of open space in the form of squares and public parks. Yet even this, given the appallingly congested and unhealthy quality of Parisian working-class housing at that time, represented a great achievement.

In Germany too the garden city movement took early root, and there it produced some interesting results. The most notable was in the city of Frankfurt am Main, where in the 1920s the city planner and architect Ernst May (1886–1970) developed a series of satellite towns (*Trabantenstädte*) on open land outside the built-up limits, and separated from the city proper by a green belt. These were not true garden cities on the Howard model, for most workers had to commute into the city – in that respect they somewhat resembled the Wythenshawe development in Manchester at about the same time – but they were remarkable for their detailed design treatment, in which May combined uncompromising use of the then new functional style of architecture with a free use of low-rise blocks, all set in a park landscape. Though the original form of the satellite towns has almost been lost in the growth of the city since the Second World War, this detailed design survives to impress the contemporary visitor. Only Berlin, where Martin Wagner was city planner, achieved the same consistently high level of design in its satellite cities during those years.

Le Corbusier

The Swiss-born architect Charles-Édouard Jeanneret-Gris, who early in his professional career adopted the pseudonym Le Corbusier (1887–1965), stands with Frank Lloyd Wright, Walter Gropius and Mies van der Rohe as one of the creators of the modern movement in architecture; and among the general public his fame is probably greater than any of the others. Yet though his best-known achievements consist of an astonishing range of individual buildings all stamped indelibly with his personality, from the Villa Savoye at Passy (1929–30) to the chapel of Notre Dame en Haut at Ronchamp, near Belfort (1950–53), his most outstanding contribution as a thinker and writer was as an urban planner on the grand scale. Of the scores of designs which Corbusier produced for city reconstructions, or for new settlements – both in France, where he worked all his professional life, and widely across the world – few materialized. The most notable are his Unité d'Habitation (1946–52) at Marseilles in France, and his grand project for the capital city of the Punjab at Chandigarh (1950–7), which is being finished only long after his death.

His central ideas on planning are contained in two important books, *The City of Tomorrow* (1922) and *The Radiant City* (*La Ville radieuse*, 1933), which is available in English translation. Unfortunately, Corbusier does not translate or summarize easily. The words pour out in no particular logical order, accompanied by diagrams which often contain the real sense of what is being said; the books seem to consist of collections of papers put together on no consistent principle; the style is highly rhetorical, and often even declamatory. But in so far as it is possible to make a very summary digest, his ideas seem to reduce themselves into a small number of propositions.

The first was that the traditional city has become functionally obsolete, due to increasing size and increasing congestion at the centre. As the urban mass grew through concentric additions, more and more strain was placed on the communications of the innermost areas, above all the central business district, which had the greatest accessibility and where all businesses wanted to be. Corbusier's classic instance, often quoted, was Manhattan Island with its skyscrapers and its congestion.

The second was the paradox that the congestion could be cured by increasing the density. There was a key to this, of course: the density was to be increased at one scale of analysis, but decreased at another. Locally, there would be very high densities in the form of massive, tall structures; but around each of these a very high proportion of the available ground space – Corbusier advocated 95 per cent – could and should be left open. The landscape he advocated, which can be seen in countless of his wrings from his Paris plan of 1922 onwards, consists, therefore, of skyscrapers separated by very large areas of intervening open space. Thus Corbusier is able to achieve the feat of very high overall densities – with up to 1,000 people to the net residential acre and more – while leaving the bulk of the ground unbuilt on (Figure 3.10 and Plate 3.4).

Figure 3.10 The Radiant City *(La Ville radieuse)*. Le Corbusier, the Swiss-French architect and planner, developed during the 1920s and 1930s the idea of a city with very high local concentrations of population in tall buildings, which would allow most of the ground space to be left open. His ideas proved very influential for a whole generation of planners after the Second World War.

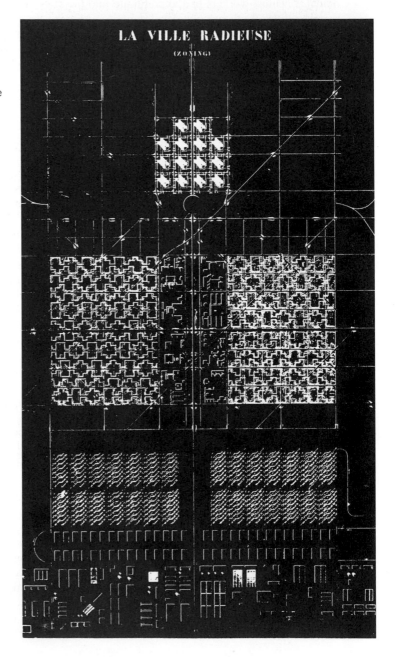

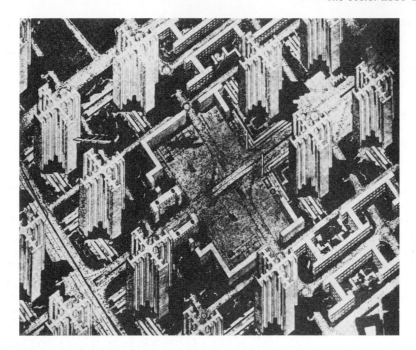

Plate 3.4 *La Ville radieuse*, as seen from the air. This is Corbusier's own imaginative conception of his radiant city. The cruciform tower blocks are designed to admit maximum light to the apartments. Dense flows of traffic on the motorway-style roads are handled by complex interchanges.

The third proposition concerned the distribution of densities within the city. Traditionally, as we noted in Chapter 2, densities of residential population are higher in the centre of the city than at the edge. Since the development of mass urban transportation from the 1860s onwards, the 'density gradient' has flattened somewhat, with lower densities at the centres and rather higher densities farther out than the rural densities which used to obtain; but it is quite noticeable nevertheless, and in Continental European cities (as well as some American cities, such as New York) it is much more pronounced than in Britain. Furthermore, there is an even more pronounced gradient of employment density, with big surviving concentrations near the centre. Corbusier proposed to do away with all this by substituting virtually equal densities all over the city. This would reduce the pressure on the central business districts, which would in effect disappear. Flows of people would become much more even across the whole city, instead of the strong radial flows into and out of the centre which characterize cities today.

Fourth and lastly, Corbusier argued that this new urban form could accommodate a new and highly efficient urban transportation system, incorporating both rail lines and completely segregated elevated motorways, running above ground level, though, of course, below the levels at which most people lived. Corbusier even claimed to have invented, in the early 1920s, the multi-level free-flow highway interchange, long before such structures were built in Los Angeles or elsewhere (Plate 3.4).

To yield the full promised results, and thus to be open to testing in practice, Corbusier's plans would need to have been applied on a very wide scale. His own diagrams show large areas of Paris, including historic quarters, razed to accommodate the new forms. This is one good reason why it proved so difficult to execute any of his ideas – especially in interwar France, where the pace of physical construction was extremely sluggish – and his notions about density have seldom been applied anywhere in the extreme he suggested. Corbusier himself became increasingly frustrated by his failure to get his plans implemented, and he began to call for an autocrat like Louis XIV or Napoleon III, who would have the boldness to execute his ideas. Nevertheless, in planning cities after the Second World War, Corbusier's general influence has been incalculable. A whole generation of architects and planners, trained in the 1930s and then from 1945 onwards, came to revere the writings of 'Corbu'; and in practice afterwards they tried to apply his ideas to local conditions. In England, for instance, his influence was particularly strong in the famous London County Council Architect's Department during the 1950s, at a time when it produced much of its best work: the celebrated Alton West estate at Roehampton in south-west London (1959), with multi-storey blocks set among areas of finely landscaped parkland, is completely Corbusian in concept (see Plate 3.5). All over Britain the remarkable change in the urban landscape during the late 1950s and the 1960s – as slum clearance and urban renewal produced a sudden unprecedented crop of skyscrapers – is a mute tribute to Corbusier's influence. Whether it was good or ill, later generations will have to decide. Certainly, by the end of the 1960s there was an increasing volume of protest at the inhumanity of the new high blocks; and it seemed doubtful whether many more would be built. Many critics were going further, and questioning the whole philosophy of massive urban renewal which was essential to the realization of Corbusier's ideas.

Corbusier has, however, had another, more subtle, influence. Though many of his ideas were intuitive rather than scientifically exact, he did teach planners in general the importance of scale in analysis. The notion that densities could be varied locally, to produce very different results while maintaining the overall density unaltered, was a very simple yet at the same time elusive one, which few grasped fully before he demonstrated it. Equally important was his insistence on the elementary truth that dense local concentrations of people helped support a viable, frequent mass-transportation

Plate 3.5 Roehampton, south-west London; the practical application of Corbusier's ideas after the Second World War by the architects of the old London County Council in their celebrated Alton West estate (late 1950s).

system. This realization, for instance, has been extremely important in the much-admired Stockholm suburbs built in the post-1950 period, where densities are systematically higher around the new underground railway stations than they are farther away.

But in general the basic difference between Anglo-American and Continental European traditions has persisted. The two lines have intermingled more in the post-1945 period than ever before, it is true: in many urban renewal areas of British cities it would really be difficult at first glance to tell whether one was in Birmingham (or Newcastle), Amsterdam, Milan or Warsaw. For many bourgeois home buyers and even some planners on the Continent there has been enthusiasm for the English idea of single-family home living and the creation of new communities in the countryside. Nevertheless, the majority of British people still appear to prefer a single-family home with garden if given the choice, while many people in Continental countries are quite firmly wedded to the advantages of inner-city apartment living.

A verdict on the seers

It may seem difficult, on first impression, to pass a general verdict on a group of planners as varied as those considered in this chapter. But in the light of the distinctions made in Chapter 1 of this book, it is possible to draw some conclusions that apply to almost all of them.

The first point is that most of these planners were concerned with the production of blueprints, or statements of the future end-state of the city (or the region) as they desired to see it: in most cases they were far less concerned with planning as a continuous process which had to accommodate subtle and changing forces in the outside world. Their vision seems to have been that of the planner as omniscient ruler, who should create new settlement forms, and perhaps also destroy the old, without interference or question. The complexities of planning in a mixed economy where private interests will initiate much of the development that actually occurs, or in a participatory democracy where individuals and groups have their own, often contradictory, notions of what should happen – all these are absent from the writings of most of these pioneers.

Howard and Geddes are, perhaps, honourable exceptions to most of this criticism. Howard's idea may have seemed utopian, but he never avoided the practical details of how to bring it about. Geddes, even more, was explicitly concerned that planning should start with the world as it is, and that it should try to work with trends in the economy and society, rather than impose its own arbitrary vision of the world. It is perhaps significant that his intellectual background was different from many of the others. An architect, by definition, starts thinking in terms of the structures he would like to build; a biologist turned geographer and sociologist starts by thinking about the nature of the society and the land s/he is planning for.

This leads us to a second point about most of the pioneers. Their blueprints seldom admitted of alternatives. There was one true vision of the future world as it ought to be, and each of them saw himself as its prophet. This is understandable, because these men were visionaries trying to be heard in a sceptical and sometimes hostile world. But if the idea is too persuasive, there is an evident risk of stifling orthodoxy.

One last point will be very evident. These pioneers were very much physical planners. They saw the problems of society and of the economy in physical terms, with a physical or spatial solution in terms of particular arrangement of bricks and mortar, steel and concrete on the ground. This again is understandable; they were trained to think in this way and their concerns were with physical development. Nevertheless, this attitude carries with it a real peril: that such planners, and those they teach and influence, will come to see all problems of cities and regions as capable of solution in these terms and only these terms. According to this view, problems of social malaise in the city will be met by building a new environment to replace the old – whereupon poor health, inadequate education, badly balanced diets, marital discord and juvenile delinquency will all go away. Similarly, problems of circulation and traffic congestion in the city will be dealt with by designing a radical new system as part of a new urban form – whereupon, of course, the problems will disappear. The notion that not all problems are capable of simple solution in these physical terms – or the more disturbing notion that there might be cheaper or better solutions to the problems, of a non-physical character – is not often found in the writings of the pioneers of planning thought we have been discussing. Nor, it should be noted, is it often found in the plans of many of those countless planners these men have influenced and inspired. The seers have made their mark as much by their limitations as by their positive qualities – striking though these latter may have been.

Further reading

Important background will be found in the works of Ashworth, Benevolo and Bell, already quoted in the reading for Chapter 2.

The best general treatment of several of the writers and thinkers discussed here is in John Tetlow and Anthony Goss, *Homes, Towns and Traffic* (Faber, second edition, 1968), especially Chapter 2. Also useful is Thomas A. Reiner, *The Place of the Ideal Community in Urban Planning* (University of Pennsylvania Press, 1963).

On (or by) particular writers, the following are important: Ebenezer Howard, *Garden Cities of To-morrow* (with preface by Frederic J. Osborn and introduction by Lewis Mumford, Faber, 1946; paperback edition, 1965); Robert Beevers, *The Garden City Utopia: A Critical Biography of Ebenezer Howard* (Macmillan, 1987); Frank Jackson, *Sir Raymond Unwin: Architect, Planner and Visionary* (Zwemmer, 1985); Walter Creese, *The Search for Environment: The Garden City Before and After* (New Haven and London: Yale University Press, second edition, 1992) on Howard, Unwin and Parker; Patrick Geddes, *Cities in Evolution* (Benn, 1968); Marshall Stalley, *Patrick Geddes: Spokesman for Man and the Environment* (Rutgers University Press, 1972; with a reprint of most of *Cities in Evolution*); Helen Meller, *Patrick Geddes: Social Evolutionist and City Planner* (Routledge, 1990); and Le Corbusier, *The Radiant City* (Faber, 1967; English translation of *La Ville Radieuse*). Robert Fishman, *Urban Utopias in the Twentieth Century: Ebenezer Howard, Frank Lloyd Wright and Le Corbusier* (Basic Books, 1977) is a useful comparative study.

John Friedmann and Clive Weaver, *Territory and Function* (Edward Arnold, 1979), contains a very useful account of planning ideas. Anthony Sutcliffe, *Towards the Planned City* (Blackwell, 1981), is a useful review of the period 1780–1914. Gordon Cherry, *Pioneers in British Town Planning* (Architectural Press, 1981), has useful accounts of Unwin, Geddes, Osborn and others. Peter Hall, *Cities of Tomorrow* (Blackwell, third edition, 2002) deals with the pioneers in some detail.

4

The creation of the postwar planning machine, from 1940 to 1952

In Chapters 2 and 3 we concentrated throughout on planning on the urban scale. But in looking at the writings of Howard, Geddes and Abercrombie we saw that, increasingly from 1900 to 1940, the more perceptive thinkers came to recognize that effective urban planning necessitated planning on a larger than urban scale – the scale of the city and its surrounding rural hinterland, or even several cities forming a conurbation and their common overlapping hinterlands. Here, the development of the idea of regional planning, in one commonly used sense of the expression, begins.

The emergence of the 'regional problem'

The difficulty – it is elementary but quite serious – is that there is another common meaning of the term 'regional planning' in modern usage. This other meaning only assumed prominence during the 1930s, as the result of the great economic depression which so seriously affected virtually all nations of the Western (non-communist) world. It refers specifically to economically planning with a view to the development of regions which, for one reason or another, are suffering serious economic problems, as demonstrated by indices such as high unemployment or low incomes in relation to the rest of the nation. Though it has some clear interrelationships with the other meaning of 'regional planning', it really represents a different kind of problem, demanding a different expertise. And commonly, the 'region' referred to in this other sort of planning is quite differently designed, and is of a different size, from the 'region' of the city-region planners.

This distinction will be discussed in more detail in a later chapter, but one simple illustration can be given here. Within Britain the northernmost part of England – including the Northumberland and Durham coalfields and the Tyne and Tees estuaries, as well as the northern Pennine uplands and the isolated industrial area of West Cumberland – has presented economic problems ever since the interwar period. The former basic industries of coal mining, shipbuilding and heavy engineering have declined; heavy unemployment and low incomes were the result in the 1930s, and during the postwar period large parts of the area, or the whole of it, have been designated a development area or development district. The unit appropriate to analysing these problems, and to providing solutions, is a fairly large one; minimally, most people would agree that the Northumberland–Durham industrial area would require treating as a single unit, and many would accept that the smaller populations in the Pennines and West Cumberland should be included for the sake of convenience, as has in fact occurred in the Northern Region Planning Council and Board. But such a unit contains quite a number of separate city regions, or cities and conurbations with their surrounding hinterlands: Tyneside,

Sunderland, Durham City, Darlington, Teesside and Carlisle, to name a few. The unit appropriate for planning of one sort may not be at all appropriate for regional planning of the other sort. And, of course, the responsibilities of the two sorts of organization will be quite different.

It is, therefore, merely confusing to give them the same name. Elsewhere, we have proposed two terms which resolve the ambiguity. The larger-scale, economic-development type of planning can best be called national/regional planning because essentially it relates the development of each region to the progress of the national economy. And the smaller-scale, physical type of planning can conveniently be called regional/local planning because it attempts to relate the whole of an urban region to developments within each local part of it.

The need for regional/local planning, as we have seen, was already coming to be recognized when Geddes was writing in 1915. But the need for national/regional planning only became fully evident in the aftermath of the Great Depression of 1929–32, and this helped trigger a series of events that, cumulatively, created Britain's post-Second World War planning system.

As the country began to emerge from the trough of the depression during 1932–6, observers noticed that certain regions which had been among those worst hit were not recovering at the same speed as the rest. These were the older industrial areas, created during the Industrial Revolution and each specializing in a narrow range of products: ships and heavy engineering in Clydeside; coal, iron and steel, ships and heavy engineering in north-east England; cotton and engineering in Lancashire; export coal, and iron and steel in south Wales. These industries, it was clear, had become extremely vulnerable to changes in the world economy: to weakening of demand from primary producing countries for industrial goods; to technological substitution (oil for coal, synthetic fibres for cotton); and to the rise of new, competing industrial powers (textiles, for instance, in India and Japan). True, at the same time new industries were growing rapidly both in production and employment. The trouble was that they were growing in quite different locations from those of old. The so-called 'new industries', representing twentieth-century rather than nineteenth-century technology (or, as Geddes had put it in 1915, 'neotechnic' as opposed to 'palaeotechnic' industry) – electrical engineering, motor vehicles, aircraft, precision engineering, pharmaceuticals, processed foodstuffs, rubber, cement and a host of others – grew rapidly in and around London, in towns like Slough and in the West Midlands (Birmingham and the associated conurbation) and East Midlands (Leicester, Nottingham, Derby and the area around). They hardly implanted themselves at all in the areas farther north, where the staple industries were dying.

The result was predictable: it was a growing discrepancy between the prosperity of the South and Midlands, and the continuing depression in the North, Wales and Scotland (Plates 4.1 and 4.2). Unemployment, 16.8 per cent in Great Britain among insured persons in 1934, was 53.5 per cent in Bishop Auckland and over 60 per cent in parts of Glamorgan; in London it was only 9.6 per cent. Despite large-scale migration from the depressed areas – 160,000 left South Wales and 130,000 left the North East during the years 1931–9 – unemployment rates remained stubbornly high in those areas right through to the outbreak of war.

The Barlow Commission and its report, 1937–40

In 1934 growing realization of the problem compelled the government to take action: the depressed areas were designated 'special areas', and commissioners were appointed for them – one for England, one for Scotland and one for Wales – with powers to spend

Plate 4.1 Jarrow in the 1930s. Jarrow became known as 'the town that was murdered' after closure of its Tyne shipyard threw nearly half a million workers into unemployment in the early 1930s. Its plight contrasted strongly with the prosperity of towns like Slough (see Plate 4.2).

public money to help invigorate the economy. 'Trading estates', on the model of the successful private enterprise example at Slough, were established in the Team Valley on Tyneside and at Treforest, near Pontypridd in south Wales. But pressure grew for a more comprehensive attack on the problem and eventually the government was prompted into action. In 1937 it appointed a Royal Commission on the Geographical Distribution of the Industrial Population under the chairmanship of Sir Anderson Montague-Barlow (1868–1951), to investigate the problem comprehensively and make recommendations.

The importance of the Barlow Commission in the history of British urban and regional planning can never be overestimated. It was directly responsible, through a chain reaction that we shall shortly trace, for the events that led up to the creation of the whole complex

Plate 4.2 Slough. An unplanned 'new town' of the 1920s and 1930s, Slough developed almost as an accident around an industrial trading estate, itself a converted wartime supply base. New industries, such as electrical goods and motor engineering, helped secure its prosperity and continued growth through the depression of the early 1930s.

postwar planning machine during the years 1945–52. Together with the name of Howard, Barlow is the most important single name in tracing the evolution of the distinctive British planning policy in the years after 1945. But together with Barlow should be coupled the name of Patrick Abercrombie – a member of the commission, a signatory of its influential minority report, and an architect (in every sense) of the postwar planning system.

In the British constitutional system the device of a royal commission permits a free-ranging, independent and deep-probing investigation of a particular problem; the commissioners need take nothing for granted. So it was with the Barlow commissioners. Their investigation was so exhaustive, and their report so authoritative and compelling in its arguments, that it actually represented a danger for later generations: the policies which were based on it became a kind of orthodoxy, very difficult to shake.

The particular contribution of the Barlow Commission to understanding and treating the problem was this: it united the national/regional problem with another problem, the physical growth of the great conurbations, and presented them as two faces of the same problem. In fact other observers had tended to do this before them, and indeed, the coupling of the two problems was explicit in the commission's terms of reference. These were, first, to inquire into the causes of the geographical distribution of industry and population, and possible changes in the causatory factors in the future; second, to consider the disadvantages – social, economic and strategic – of the concentration of industry and population into large centres; and third, to report on remedies that were necessary in the national interest. Two things should be noted: first, that the national/regional distribution of industry and people was linked to the question of the concentration of population within regions – a rather different question; second, that the terms of reference were deliberately loaded, since it was assumed that disadvantages existed and that, implicitly, they far outweighed any possible advantages.

Given these terms, the findings of the commission were perhaps predictable. On the first point, the report, when it emerged in 1940, confirmed the general impression that the growth of industry and population during the interwar period had been strikingly concentrated in the prosperous areas of the South and Midlands, and above all, around London. Table 4.1, taken from the commission's analysis, shows that only in two areas of the country, London–Home Counties and the Midlands, was the growth of the insured employment greater in this period than in the nation as a whole; in London–Home Countries it was nearly double the national average. Another analysis of the same figures, also in Table 4.1, is perhaps even more striking: it shows that London and the Home Counties accounted for over two fifths of the growth of insured employment in this period, though they had less than one quarter of the employment at the start of it. Geographers testifying before the commission argued for the existence of a main industrial axis, or 'coffin' area (the name referred to its shape) embracing the London region, the

Table 4.1 *Insured workers, 1923 and 1937*

	Insured workers (thousands)		Per cent increase	Per cent of national increase
	1923	1937	1923–37	1923–37
London and the Home Counties	2,421	3,453	42.6	42.7
Staffordshire, Warwickshire, Worcestershire, Leicestershire, Northamptonshire	1,212	1,554	28.2	14.1
Lancashire	1,697	1,826	7.6	5.3
West Riding, Nottinghamshire, Derbyshire	1,403	1,614	15.0	8.7
Northumberland and Durham	619	648	4.7	1.2
Mid-Scotland	792	868	9.6	3.1
Glamorgan and Monmouth	457	437	−4.4	−0.8
Rest of Great Britain	2,225	2,844	27.8	25.6
TOTAL	10,826	13,244	22.3	100.0

Source: Barlow Report

Midlands, Lancashire and Yorkshire, into which industry and people were concentrating; the commissioners found that this was not a very helpful framework of analysis because so much of the growth was at the southern end of the belt, and virtually none of it at the northern end.

What were the causes of these trends? The Barlow Report found that the pattern of industrial growth – or the lack of it – was dominated by what has come to be called the 'structural effect'. This very important term – it will be discussed more fully in a later chapter – refers to the finding that the growth of the more prosperous areas can almost wholly be explained in terms of their more favourable industrial structure. In other words, their regional economy was so dominated by growth industries that by applying the national growth rate for these industries it was possible to predict the growth of the region; such industries were not, in most cases, expanding faster in these regions than anywhere else. For the depressed areas the conclusion was further depressing: their basic industries were declining so fast that they were running the whole economy downhill – it would be necessary to make superhuman efforts just to keep the economy in the same place.

Barlow's analysis of the causes of this locational pattern is still a classic: it has already been referred to, but it deserves a longer summary. Nineteenth-century industry, the analysis ran, had been diverted towards fuel and raw-material supplies and to navigable water, but twentieth-century industry needed these factors much less: their pull being weakened, industry would naturally gravitate to its main markets. But the market, that word used so casually in accounting for industrial location, is actually a complex thing: it includes sales to other industries, export agencies and a host of special sales facilities; and all these tend to be located in very big population centres. Such areas also tend to have a wide range of different labour skills and specialized services, which smaller industrial towns lack. Yet this pattern of forces, if it continued, would pull new industry away from the coalfields, which tended to be distant from the main marketing centres, leaving large concentrations of population and social capital stranded there. The Barlow Commission could find no good cause why the pattern of forces, left to itself, should start working in a different direction. So the question was: was there any reason for taking action to modify the natural course of events?

This led the Barlow Commission naturally to the second of the terms of reference: the analysis of disadvantages. Here the commissioners were led into quite new and uncharted territory. Hardly anyone, anywhere in the world, had systematically considered questions like this before. And it should be remembered that then, very few economists were interested in urban affairs: there was no body of theory, no empirical research, to help the commissioners. They looked systematically, and in detail, at records of public health, at housing, at traffic congestion, at the patterns of journeys to work, at land and property values. Then, *in camera*, sitting under the threat of imminent war, they heard the evidence of defence experts on the strategic dangers from bombing attacks on big cities – dangers which proved only too true in countless cases during the Second World War.

Some of the resulting analysis has been outdated by subsequent social changes; some of it, indeed, was tendentious and inconclusive at the time. For instance, the commission concluded that, broadly, housing and public-health conditions tended to be worse in big cities (and in conurbations) than in small towns. But even then the evidence for that was contradictory: London, for instance, had better public health records than the national average. And since then the position has changed beyond recognition. As a result of general improvements in public health – better maternal and infant care, free national health facilities, higher real incomes – indices like infant mortality have greatly improved, and the differences between one part of the country and another have been replaced.

Some of the indices for poorer health, and for overcrowding of homes, are found in small towns. In an age when a nuclear holocaust could mean the virtual end of civilization, the strategic arguments against big cities have less force. Other arguments, though, continue to have force – sometimes, even greater force. Journeys to work have lengthened, though perhaps not so much if the measure used is time rather than miles; traffic congestion may have worsened (though it is very difficult to make comparisons over a long period, and some evidence for London indicates that the traffic actually speeded up between the 1930s and the 1960s, though it has subsequently slowed); land and property values have certainly escalated, especially near the centres of the biggest cities; some of the most serious housing problems, including homelessness, are certainly concentrated in the inner areas of the conurbations.

In the twenty-first century a royal commission would doubtless try to fit all this information within a theoretical economic framework, and to produce a cost–benefit analysis (see Chapter 9) of the advantages versus the disadvantages of life in big conurbations, all fully quantified in money terms for the sake of comparison. Such techniques were not open to Barlow, and indeed some would argue that they can be positively misleading, by giving a spurious impression of exactness. What the Barlow commissioners did was to sift the evidence as best they could. They concluded that the disadvantages in many, if not most, of the great urban concentrations far outweighed any advantages and demanded specific government remedies. London, they thought, represented a particularly urgent problem which needed special attention.

So far the commissioners were agreed. When they came to discuss remedies – the third part of their terms of reference – they split. It was clear to all that since no democratic government could direct people where to live, the controls would have to be applied to the location of new industry. In the conditions of the late 1930s such controls on the freedom of industry were considered radical, and even revolutionary. So the commissioners split into two groups. The more moderate majority suggested that in the first place there should be controls only on the location of new industry in and around London, to be imposed by a board. The more radical minority – including the influential Professor Abercrombie – recommended more general controls on the location of industry throughout the whole country, to be administered by a new government department set up for the purpose.

In the event, as we shall see, when the government came to act on the Barlow recommendations – in 1945 – they opted for a modified version of the radical variant. But in addition to this central investigation and set of recommendations the Barlow commissioners also studied a number of important related problems. Among these were the technical problems of controlling the physical growth of cities and conurbations, and of preserving agricultural land, through the establishment of a more effective system of town and country planning; and the linked problem of compensation and betterment in planning. On neither of these two questions could they reach definite recommendations: each was so complex, they concluded, that it needed further expert study. Similarly, though the commission endorsed the general idea of building garden cities, or new towns, in association with controls on the growth of the conurbations, they thought that further investigation was needed of the ways in which this should be done.

The aftermath of Barlow

The Barlow Report was submitted to the government at the outbreak of war and was actually published in the middle of the so-called phoney-war period, a few months before Dunkirk, in February 1940. Shortly afterwards the war effort fully engaged most people's attention. But at the same time, in a remarkable mood of self-confidence about the future, the wartime government embarked on the follow-up studies which the Barlow Report had recommended.

The result was a remarkably concentrated burst of committee work and report writing, from 1941 to 1947. A whole succession of official reports, either from committees of experts or from planning teams, made recommendations to government on various specialized aspects of planning. These reports, known commonly after their chairman or team leader – Scott, Uthwatt, Abercrombie, Reith, Dower, Hobhouse – laid the foundations of the postwar urban and regional planning system in Britain. Then, in an equally remarkable burst of legislative activity from 1945 to 1952, postwar governments acted on the recommendations; not always following them in detail, they nevertheless enacted them in essence. A series of Acts – the Distribution of Industry Act 1945, the New Towns Act 1946, the Town and Country Planning Act 1945, the National Parks and Access to the Countryside Act 1949, and the Town Development Act 1952 – created the postwar planning system. Though since modified in many respects, its broad outlines have survived.

In this chapter we shall consider in sequence first the principal reports which provided the foundations of the system; then the legislation which brought it into being. Finally, we shall try to sum up the essential character of the system: its positive values and its limitations.

The Foundation Reports

Scott and Uthwatt

The first of the studies, the report of the Committee on Land Utilization in Rural Areas, was published in 1942. Though this committee is known after the name of its chairman, Sir Leslie Scott (1869–1950), the report bears the unmistakable imprint of the vice-chairman and chief author, the geographer Sir Laurence Dudley Stamp (1898–1966). The burden of this report was that good agricultural land represented a literally priceless asset: unlike most other factors of production, once lost it was lost for good. Therefore, the report argued, the community should set up a planning system embracing the countryside as well as the town; and this system should regard it as a first duty to preserve agricultural land. In the case of first-class land – which Stamp's own land utilization survey in the 1930s had shown to be a very small part of the total land area of Britain (about 4 per cent) – there would be an automatic and invariable embargo on new development; this would prevent any recurrence of the process whereby west London expanded over the fertile market-garden lands of Middlesex. But even elsewhere the Scott report suggested the principle of the 'onus of proof': wherever development was proposed, it should be for the developer to show cause why the proposed scheme was in the public interest. Otherwise the existing rural land use should have the benefit of the doubt.

It is easy to see the attraction of such an argument in 1942, when the blockade on the seas was making Britain more dependent on home foodstuffs than at any time since the early nineteenth century, and when British farmers made heroic efforts to increase production of basic cereals. And though the onus-of-proof rule has never been applied so rigidly in actual postwar planning, there is no doubt that the general sentiment behind the case has been very powerful in supporting the notions of urban containment and of encouraging higher-density urban development so as to save precious rural land. What is interesting is that even in 1942 the voice of an economist was heard to attack this view as lacking in economic sense. Stanley Dennison, a member of the committee, signed a minority report which suggested that the true criterion should be the value of the land to the community in different uses. In fact, Dennison was really calling for the application of cost–benefit analysis to urban planning decisions – a technique then hardly understood anywhere. Naturally, his voice went largely unheeded.

The Expert Committee on Compensation and Betterment, whose final report of 1942 is generally know after its chairman, Sir Augustus Andrewes Uthwatt (Lord Justice Uthwatt, 1879–1949), dealt with a perennial problem of urban development whose origins in English law can be traced back to 1427. It has two aspects, which are linked. The first is the problem of compensation: when a public body has to buy land compulsorily, for a new highway or a new school, for instance, what is the just rate of compensation to the dispossessed owner? At first, the answer might seem simple: the public body should pay the current market value, since that will make the owner no worse off, nor better off, than if he or she sold in the market; furthermore, the public body ought to want the land enough to be willing to pay the going market price. But the complication is that the public body, unlike most private buyers, may have helped to create a large part of the land value it has to pay for. If, for instance, it announces a new motorway, land values might arise around the likely position of an interchange with the existing main road; if the community then had to pay this enhanced value, it would seem unfair.

The complication described here has a name in law: betterment. Originally, this term was reserved for the case where the community took action which clearly made some people better off; the legal argument was that the community should then be able to claim a special tax from these people, reflecting the fact. (The 1427 case, mentioned above, referred to sea-defence works.) But then, it was seen that public actions may be more subtle, and yet make people better or worse off. Suppose the community takes the power to stop building on a fine piece of countryside. Some people – those who owned the land – will be worse off because they cannot enjoy the profit from development. Others – those who lived next to the area – will be directly better off, because they now have an unimpeded view which they expected to lose; they can now sell their land at a profit. Yet others – the general public who can come and enjoy the scenery – are indirectly better off. The third group are difficult to deal with, except perhaps by imposing a charge for entry to the area (as is done, occasionally, in park areas). But with regard to the others, it would seem that in fairness the community should pay compensation to the first group and claim betterment from the others.

The Uthwatt Committee report went in great detail into the conceptual and technical aspects of this problem. Finally, they concluded that the complexities were such that the community would do best by a fairly simple, crude approach: cutting the Gordian knot, as the committee described it. Land which was not developed – that is, all the rural land of the country – should in effect be nationalized: the state should acquire it, paying compensation to the owners on the basis of the value at some historic date in the recent past. But for the time being, and in some cases perhaps for all time – until such time as the land was needed for urban development – the owner could remain on the land. So his/her compensation would be limited to the loss of his/her right to develop the land. If and when the state needed the land for building, it would pay him/her additional compensation for expropriating him/her altogether. Then it could sell or lease the land to a developer. Within the built-up areas, on the other hand, the committee recommended that any redevelopment of existing property should be carried out by the local authority, who would buy the land on the basis of its value at some recent date and carry out the redevelopment itself. Lastly, the committee proposed that all property owners should pay a regular betterment levy, calculated at the rate of 75 per cent of the increase of the value of the site alone (without the building) since the previous valuation; for this purpose, in addition to the usual valuation of property for rating, there would need to be a separate regular valuation of the site alone.

The Uthwatt Committee solution truly cut the Gordian knot; and it could have been effective. Its most important feature was that in relation to the major problem – the development of rural land for urban purposes for the first time – it did not need the land market to work at all in the old way; the state would be the sole buyer. This was its technical strength, but its political weakness. Though the report was prepared by disinterested land experts, not by left-wing politicians, it generated immense controversy and opposition on the ground that it advocated land nationalization. The coalition government of the time at first took no action on it, but then (in 1944) announced that after the war it would reform the law on another, less radical basis. But in the same year, in a Planning Act, it did provide an expedited procedure which allowed the blitzed cities to buy land for reconstruction on a quick and cheap basis.

Abercrombie and Reith

In 1944 the wartime government received – and early in 1945 it published – another major report: Patrick Abercrombie's Greater London Plan. Starting boldly from the position that the Barlow recommendations on industrial location controls would be accepted and acted upon, and that population growth in the country as a whole would be negligible – an assumption which corresponded to the best demographic forecasts of the time – Abercrombie worked on the basis that the population of London and its surrounding ring, a wide area stretching roughly 30 miles (50 kilometres) in any direction from London, could be held constant. The task he set himself was to achieve a massive decentralization of people from the inner, more congested part of this vast region to the outer rings. Within the inner part, the County of London Plan (on which Abercrombie had cooperated) had demonstrated that if the slum and blighted areas were to be redeveloped to adequate standards of open space, a planned overspill programme for over 600,000 people would be needed; outside the LCC area, Abercrombie now calculated that the corresponding overspill would amount to an additional 400,000, giving over 1 million in all. Up to 1939 the accommodation of these people would have been carried out in the most obvious way: by building peripheral estates at the edge of the conurbation, thus adding further to the urban sprawl. Following the Barlow recommendations to the letter, Abercrombie proposed to end all this by a bold device. A green belt would be thrown around London, at the point where the conurbation happened to have stopped at the outbreak of war in 1939; five miles wide, on average, it would provide an effective barrier to growth and also act as a valuable recreational tract for Londoners (Figure 4.1).

Most importantly, the very width of the belt would fundamentally affect the treatment of the overspill problem. If the overspill were removed to the outer edge of the green belt, or even farther, that would put it well beyond the normal outer limit of commuting to London at that time. New communities could then be created to receive these 1 million people, which would be truly what Ebenezer Howard had intended: self-contained communities for living and working. Abercrombie thus seized the unique opportunity that had been offered to him: to produce a total regional plan as Geddes had advocated, and thus to carry out the principles that Ebenezer Howard had established nearly half a century before. Abercrombie, therefore, proposed that about 400,000 people be accommodated in eight more or less completely new towns with an average size of about 50,000 each, to be built between 20 and 35 miles (35 and 60 kilometres) from London; another 600,000 should go to expansions of existing small country towns, mainly between 30 and 50 miles (50 and 80 kilometres) from London, but some even more distant than this.

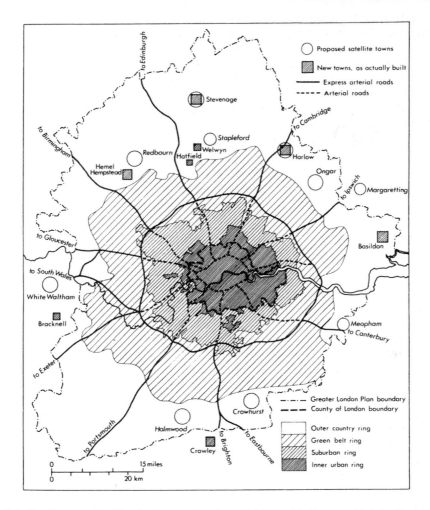

Figure 4.1 The Abercrombie Plan for Greater London, 1944. Patrick Abercrombie's bold regional plan involved the planned dispersal of over a million Londoners from the congested inner urban ring, across the new green belt, which would limit the further growth of the conurbation into planned satellite towns – the famous London new towns.

All the other major conurbations of Britain were the subject of similar wide-ranging regional plans, either at the end of the war or shortly afterwards: Abercrombie himself prepared two of them, one for Glasgow and one (jointly) for the West Midlands. All made radical and far-reaching proposals for planned urban decentralization on the principles advocated by Ebenezer Howard, though in none, of course, was the scope so large as in London. Clearly, such large-scale population movements – to be carried out in a relatively short time – posed major problems of organization; the existing structure of local government appeared completely unsuited to deal with them. And apart from the experimental private new towns at Letchworth and Welwyn and the municipal venture by Manchester at Wythenshawe, there was no experience in building new towns.

Therefore, just after the end of the war, the incoming Labour government commissioned yet another major committee report. Dispelling some doubts on the matter, it announced that it supported the principle of planned decentralization to new towns, and appointed a committee to consider ways of building them, to be headed by the redoubtable Lord Reith (1889–1971) – creator of the BBC and the first minister responsible for planning in the wartime coalition government, until his enthusiasm and intransigence caused his abrupt dismissal. With his usual energy, Reith set his Committee on New Towns to work and quickly produced two reports – the second early in 1946. It recommended that new towns should normally be built very much as Howard had proposed them, with a size range of 30,000 to 50,000 or perhaps 60,000; though it had little to say about the Social City principle which Howard thought so important. As for organization, it confirmed that the existing local government structure was not suitable for the task. The new towns, it proposed, should each be built by a special development corporation set up for the purpose, generally responsible to Parliament, but free of detailed interference in its day-to-day management, and with direct Treasury funding. The formula, in other words, was rather like that of Reith's own beloved BBC.

Almost certainly, Reith's formula was the right one. In building the new towns, freedom for managerial enterprise and energy had to be given priority over the principle of democratic accountability; if the new towns had had to account for every step to a local authority, they could never have developed with the speed they did. This was particularly so, since almost by definition the existing local community tended to be opposed to the idea of any new town at all. When the new town was largely completed, the Reith Committee argued, that would be the appropriate time to hand it over to the local community for democratic management.

Dower and Hobhouse

With the publication of the Reith Committee's second and final report in January 1946, an extraordinary burst of official committee thinking had come almost to an end. Hardly anywhere, in any nation's history, can such sustained and detailed thought have been given to a set of interrelated and highly complex problems within a single field. Only two further reports, in a separate specialized area, remained to complete the list of recommendations. The Dower Report on National Parks, a one-man set of recommendations commissioned by the government from John Dower, a well-known advocate of the establishment of a national parks system, was published in 1945; it was followed in 1947 by the Hobhouse Committee Report on National Parks Administration with detailed recommendations about the organization of the proposed parks. Both reports agreed that the parks should be speedily established in areas of outstanding scenic and recreational importance, and that they should be fully national in character; further, that they should then be positively developed for the outdoor enjoyment of the people, as well as for purposes of conservation of resources. This suggested that the parks organization should be outside the normal framework of local government, there being a parallel here with the Reith Committee recommendations for new towns. A National Parks Commission should be formed, with full executive powers to plan and supervise the work of establishing the parks; it should then devote its powers upon an executive committee in each park. The recommendations here, in fact, followed fairly closely the organization of the outstandingly successful and well-established National Parks Service in the United States; this national parks there are run by a bureau of a federal government department – the Department of the Interior.

The period of committee sittings and report writing, therefore, was concentrated into a short period between 1937–47, with the greatest activity actually in the wartime years of 1940–5. Together, shortly after the end of the war, the completed reports constituted an impressive set of blueprints for the creation of a powerful planning system. But the existence of these blueprints provided no guarantee that action would be forthcoming.

In the event, the powerful reforming mood which swept over the country at the end of the war – and which expressed itself in the surprise victory of the Labour Party in the July 1945 general election – provided the impetus to turn recommendations into legislative action. The report writing period of 1940–7 was followed, with a momentary overlap, by the legislative burst from 1945 to 1952, which we must now follow. In doing so, we shall try to establish some sort of logical order, so as to bring out the interrelationships between the different pieces of legislation. In one or two cases, this will mean important divergences from the chronological order of the different Acts of Parliament.

The legislation

The 1945 Distribution of Industry Act

The first in this great legislative series, both chronologically and in terms of the whole logical structure, was the Distribution of Industry Act of 1945, which was passed by the coalition government just before the July election. Its great importance was that it provided for comprehensive government controls over the distribution of industry, of a negative as well as a positive kind. Upon the recommendation of the Barlow minority, these extended over the whole country. In future, any new industrial plant, or any factory extension, over a certain size (which was originally fixed at 10,000 square feet (929 m^2) or 10 per cent, but which was varied somewhat subsequently) had to have an industrial development certificate (IDC) from the Board of Trade (and its successor, the Department of Trade and Industry) throughout the period since 1945 to steer industry away from London and the Midlands, and towards the former special areas.

But the Act also contained new provisions for the positive encouragement of new industry in these areas – henceforth to be known as the development areas. (In 1945, they consisted of Merseyside, north-east England, West Cumberland, central Scotland and south Wales.) Industrialists setting up plants in these areas would receive a variety of government inducements, including specially built factories, ready-built factories for occupation at low rents, investment grants for the installation of new equipment, and loans.

It seemed like an impressive combination of stick and carrot; but it contained three important limitations. The first was that the system of control applied to factory industry only; location controls were not applied to offices at all until nearly 20 years afterwards in November 1964, and they have never been applied at all to other forms of tertiary (service or non-manufacturing) industry. Probably the reason for this failure is to be found in the faulty analysis performed by the Barlow Commission; the employment figures available to the commissioners excluded a great deal of service industry, because of the incomplete national insurance coverage at that time, and so they underestimated the degree to which the rise in employment in London was the result of tertiary sector. In any event, in their critical recommendations the whole Commission – the majority and minority alike – seemed to confuse two meanings of the word 'industry': one, meaning all types of employment; the other meaning the second, limited definition; and the 1945 Act followed them. In the event, employment in manufacturing stagnated in Britain after 1945; the whole net growth of employment was in the service industries.

A second limitation – especially serious, in view of the stagnant state of factory employment – was that the incentives applied chiefly to provision of capital equipment. This meant, paradoxically, that a highly capital-intensive firm using a lot of machinery and very little labour could get generous grants to go to a development area, where it would do virtually nothing to reduce local unemployment. In fact, a firm could actually use the incentives to automate and reduce its labour force. Preposterous as this may seem, there are indications that in one or two cases it actually happened.

A third limitation was simply that the Act left many loopholes. Any firm that was frustrated in its attempts to get an IDC in London, or the Midlands, could easily do one of two things. Either it could extend its existing plant by just under 10 per cent (or 5 per cent, depending on the regulations at the time) a year, thus increasing by 50 or 100 per cent in a decade. It could supplement this by moving out warehouse or office space into separate buildings, which did not need a certificate, taking the space for factory production. Or it could simply buy a 'second-hand' vacated factory in the open market. There is plenty of evidence therefore that, though the whole policy did steer jobs to the development areas, the effect was far less spectacular than many people hoped. Above all, contrary to expectations, the Act provided no sure machinery at all for curbing the growth of employment in the South East or the Midlands.

The 1946 New Towns Act and the 1952 Town Development Act

The New Towns Act of 1946 passed into law with remarkable speed soon after the Reith Committee's final report in order to expedite the designation of the first of the new communities. The Committee's recommendations were faithfully followed. New towns were to be designated formally by the minister responsible for planning – a Ministry of Town and Country Planning had been set up in 1943, and this was one of its first important functions. The minister would then set up a development corporation, responsible for building and managing the town until its construction period was finished. The Act left open the critical question of what was to happen to each town after that date, but it was generally expected (as the Reich Committee had proposed) that it would revert to the local authority. However, in 1958 the government of the day finally decided instead to hand them over to a special statutory authority, the Commission for the New Towns. This aroused a great deal of controversy, but there was an overpowering reason for it; as Ebenezer Howard had prophesied, new town construction proved a very good investment for the community because of the new property values that were created, and it would seem inequitable to hand over these values to the local authority which happened to occupy the area. If the values belonged to the community, they belonged to the whole community. Ironically, under legislation passed in 1980, the new towns were compelled to sell off some of their most profitable assets in the form of their commercial centres. Much of their estate is, however, still managed by a public corporation, English Partnerships.

Progress was rapid after the Act was passed: the first new town, Stevenage, was designated on 11 November 1946 – the very day the Act received the royal assent (Figure 4.2 and Plate 4.3). Between 1946 and 1950, no less than 14 new towns were designated in England and Wales: eight of them around London, to serve London overspill as proposed in the Abercrombie Plan of 1944 (though not always in the locations proposed in the plan, some of which were found to be unsuitable); two in north-east England to serve the development area; one in south Wales to serve a similar purpose (though it was actually just outside the development area); two in central Scotland for the same reason (one of which also received overspill from Glasgow); and lastly one attached to a prewar steel works. Then for a decade progress virtually ceased from 1950 to 1961, only one new town – Cumbernauld in Scotland – was designated, and in 1957 there was an announcement that no more new towns would be started. But in 1961 there was an abrupt reversal of policy – for reasons which are analysed more fully in Chapter 6

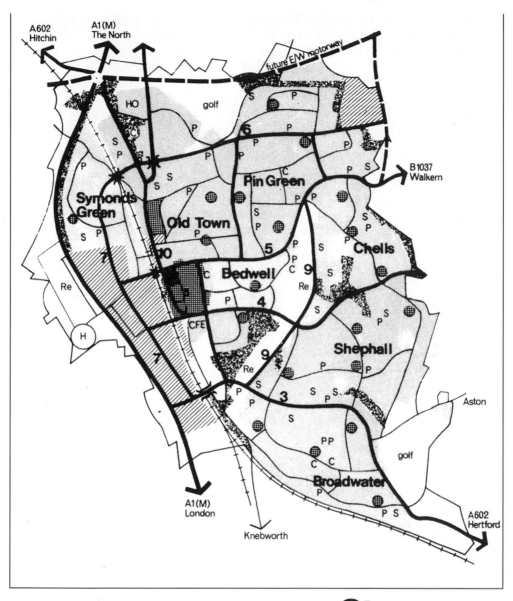

Figure 4.2 The master plan for Stevenage, the first new town to be designated (in 1946). Built for London overspill and sited 30 miles (50 kilometres) north of the metropolis, in Hertfordshire, Stevenage is a good example of the Mark 1 new town of the 1940s, designed in neighbourhood unit principles.

Plate 4.3 Stevenage town centre. The Town Square was one of the first pedestrian precincts in a British town centre. Stevenage has fared better 60 years on compared to some of the other Mark 1 town centres.

– and between then and 1970 no less than 14 further new towns were designated in Great Britain. In 1980, 34 years after the passage of the Act, Britain's 28 new towns accounted for over 1 million people, with more than 700,000 new houses built.

New towns were, however, only one arm of the policy which Abercrombie had proposed for Greater London; the other was the planned expansion of existing country towns, in order to serve the twin purposes of the development of the more remote rural areas and the reception of overspill. The notion here was fairly consistent: such towns would be more distant from the conurbations than the new towns; they would have an existing population and existing industry; and the new towns mechanism would not be suitable for their expansion. Rather, they should be aided in reaching voluntary agreements with the conurbation authorities, with a financial contribution from central government to cover necessary investments. The Town Development Act of 1952, prepared by the Labour government before the 1951 election but passed by the Conservatives after it, provided for this machinery. At that time, it was thought that the programme of new designations was substantially complete, and that further overspill could and should be provided for by the new Act. But in practice the financial inducements proved insufficient at first, and local authorities with housing problems in the conurbations found real difficulty in reaching agreements. By 1958, indeed, the whole procedure of the Act had provided a derisory total of less than 10,000 houses in England and Wales. Thereafter progress was more rapid, with some really big agreements reach by London for the large-scale expansion of Basingstoke, Andover and Swindon; and by 1977, nearly 89,000 houses in England and Wales had been constructed under the provisions of the Act. The programme by then was virtually complete, and the biggest overspill authority – Greater London – was disengaging from it.

The 1947 Town and Country Planning Act

Between them, the 1946 Act and the 1952 Act eventually provided effective mechanisms for the planned overspill of hundreds of thousands of people from the conurbations into new planned communities outside. It seems, from the statements of the time, that such planned developments were confidently expected to provide for the great majority of the whole new housing programme in the country in the postwar period. Between 1946 and 1950, the public sector – the local authorities and the new towns – built more than four in five of all new homes completed. It appears to have been thought that private speculative building for sale would never again achieve the role that it had played during the 1930s. Abercrombie, for instance, assumed in his 1944 plan that over 1 million people would move from London to new communities in planned overspill schemes, as against less than 250,000 moving by spontaneous migration. Between them, three types of public housing authority – city authorities building on slum clearance and renewal sites, new town development corporations and country towns (or the city authorities building in those towns at their invitation) – would provide for the great bulk of the people's housing needs; and these programmes would all proceed within the orderly framework of city-regional plans.

This is important, because it provides the setting within which the 1947 Town and Country Planning Act was drawn up and passed. The 1947 Act, one of the largest and most complex pieces of legislation ever passed by a British Parliament, was indeed the cornerstone of the whole planning system created after the Second World War. Without it, effective control of land use and of new development would have been impossible. Green belts, for instance, could not have been drawn around the bigger urban areas in order to contain and regulate their growth; a plan like Abercrombie's would, therefore, not have been enforceable (Figure 4.3). The effectiveness of the powers is in fact remarkable by international standards since, though many countries have powers to limit development on paper, demonstrably they do not work in practice. But in seeing how this was achieved, it is worth remembering that the system was designed to deal with only a limited part of all the new development; the rest would be carried out in planned public developments like new towns. Such a system was nevertheless necessary, in order not to compromise the fairly radical public programmes.

The first important feature of the 1947 Act, and the key to all the rest, is that it nationalized the right to develop land. This was what the Uthwatt Report had recommended in 1942, in respect of rural land; the 1947 Act extended this to all land, but it did not provide for eventual state take-over when the land was needed for development (save, of course, in the case of compulsory purchase by public authorities for their own schemes). Apart from these last, the land market was still required in order for development to take place: private owners would sell directly to private developers. The Uthwatt proposal was really more consistent with the situation which seems to have been predicted for the postwar period; in a world where the great majority of all new development was in the public sector, it was surely logical to provide for outright state purchase of the land just before development took place, in the case of public and private schemes alike. But the government drew back from this extreme step.

At any rate, the nationalization of the right to develop was the minimum necessary to ensure effective public control over the development and use of land in accordance with a plan. The second feature of the 1947 Act, therefore, was the linkage of

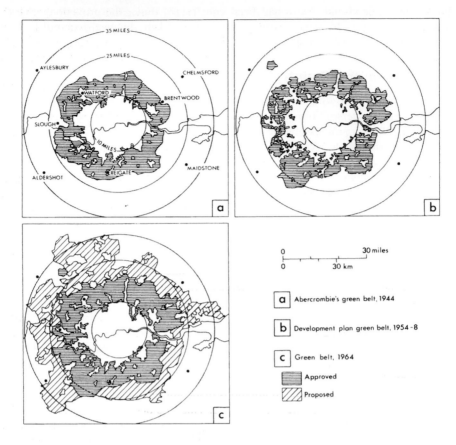

Figure 4.3 The London Green Belt, 1944–64. Earliest of the postwar green belts to be established around Britain's urban areas, the metropolitan green belt has increased since Abercrombie's original 1944 proposals. The green belt had several purposes, including urban containment, agricultural protection and the preservation of land for recreation. Almost 70 years on, the green belt remains but its purpose and form is now being questioned in the light of changing economic, environmental and social pressures.

plan-making and development control through the creation of new local planning author-ities charged with both functions. These were to be the largest available existing local authorities: the counties and the country boroughs in England and Wales, the counties and cities and large burghs in Scotland. At one step, the number of authorities responsible for planning was reduced from 1,441 to 145 in England and Wales. These authorities were charged with the responsibility of drawing up, and quinquennially revising, a development plan for their area, based on a survey and analysis (as recommended many years before by Patrick Geddes); the plan, to consist of a written statement and maps, was to show all important developments and intended changes in the use of the land over a 20-year future period. This plan was to be submitted to the minister responsible for planning, for his approval; thereafter, the local planning authority was to administer development control in accordance with the plan. Henceforth, anyone wanting to develop – the term was carefully defined in the Act, but basically meant changing the use of the land by creating structures on or in the land – must apply to the local authority for planning permission; the authority could refuse permission on the ground that the development was not in accord with the plan, or on other grounds, and though there was the right of appeal to the minister (who might order a public inquiry at his discretion) the aggrieved owner had no other legal redress. This was only possible because of the nationalization of development rights embodied in the Act; these rights were then in effect presented by the state to the local planning authorities.

A third important feature of the Act was compensation. Just as the government at that time was nationalizing coal mines and railways, paying compensation to the shareholders for their interests, so here the government provided compensation to landowners for lost development rights. Many owners, after all, might have bought land expecting to develop it, and it seemed unjust to deny them this right without due compensation. The Act, therefore, provided a formula: all the development rights in the country were to be valued and added up, and then scaled down to allow for double counting. (This arose because round a city, only a certain percentage of the available plots were likely to be developed in any one period; but naturally, all owners thought that theirs would be the lucky plot, and valued it accordingly.) Then, on a day in the future, all scaled-down claims would be paid. Thereafter, the owners of land (who would continue in possession, retaining the right to enjoy the land in its existing use) would have no further claim to development rights; the state – or the local planning authority to which it had passed them – could exercise these rights freely.

The logic of this led to the next important feature of the Act. If development rights were nationalized, and if owners were compensated for losing them, they had no further claim to enjoy financial gains from any development: if, subsequently, the local planning authority gave them permission to develop, then the community should enjoy any profits that arose. The 1947 Act therefore provided that in the case of permission to develop being granted, owners should pay the state a 'development charge', representing the monetary gain arising; and under regulations made afterwards under the Act, this charge was fixed at 100 per cent of the gain in value. This was perfectly logical and equitable; the only difficulty was that it did not work. We saw above that in the Act, the government shied away from the radical Uthwatt solution of actually taking the land needed for development; in consequence, the private land market was still required to work. Yet the 100 per cent charge removed all incentive for it to work. By 1951 there was evidence that, to make the market move, buyers were paying over the odds for land: they paid the development charge twice over, once to the state and once to the seller.

This was inflationary; still more inflationary would be the once-for-all payment of compensation for lost development rights, amounting to £300 million, which was due in 1954. So in 1953 the new Conservative government to all intents and purposes scrapped the financial provision of the 1947 Act. They abolished the development charge (though many argued that the right course would have been to reduce it, not cut it out altogether). And they provided that compensation would be paid only as and when owners could show that they had actually applied for permission to develop and their application had been rejected. The end of the development charge, however, created an anomaly: landowners who could get development permission would enjoy the whole of the resulting speculative profit (though they now lost their claim on the £300 million); but if their land was compulsorily purchased by public authorities, as for a road or a school, they got only the existing use value. Under the 1947 Act this was logical (since the private seller got no more than this, after s/he had paid the development charge); now it was not. So in 1959, to restore equity, the government returned to full market value as the basis for compulsory purchase by public authorities. The one exception was where – as in the important case of a new town – the authority should not pay any value which resulted from its own actions on pieces of land around the land in question. In other words, a new town development corporation must pay the owner its assumed full market value in the event that no new town was being built. This completely artificial assumption was necessary to prevent the absurdity of a new town paying values which it had itself created.

Two further attempts were made to grapple with the intractably related problems of compensation and betterment, which the financial provisions in the 1947 Act had attempted without success to resolve. The first was the 1967 Land Commission Act, passed by the 1964–70 Labour government and repealed by the Conservative administration after the 1970 election. Though discussion of the Act logically comes later on, in Chapter 6, it is useful to discuss its compensation and betterment provisions here, to see how they relate to the 1947 solution.

The 1967 Act was a partial return to the Uthwatt solution of 1942. Originally, the Labour government had thought to take the Uthwatt proposals more or less in their entirety, so that a Land Commission would be set up to buy any and all land needed for development; but they drew back from this extreme step. Instead the commission would progressively build up a land bank which it could release for development when needed. Additionally, the Act provided that a betterment levy should be charged whenever land changed hands (being payable by the seller), and at the point of development. This was similar to the levy proposed by Uthwatt, but it differed in two ways. First, it was a lower rate: 40 per cent rising to 50 per cent and perhaps more, as against Uthwatt's 75 per cent. Second and more fundamentally, it was not to be charged regularly whether or not the owner actually profited from the rise in value, as Uthwatt had proposed, but only when s/he realized the increase through sale or development. So it was a very watered-down version of the original Uthwatt idea. Nevertheless it did cut the Gordian knot – in Uthwatt's phrase – by taking some betterment for the public purse, while, it was hoped, leaving the owner with an incentive. And it did reduce the burden of land purchase for public authorities, since the Land Commission paid a sum net of levy when it bought either in the free market or compulsorily. Unfortunately, because of uncertainty whether the Act and the commission would survive, the effect seems to have been inflationary, just as after 1947: sellers paid the levy to the commission, but then added at least part of it again to the price they charged to buyers. And just as after 1951, this was one reason the incoming government gave in 1970 for rescinding the provisions altogether.

Labour tried again, passing the Community Land Act in 1975. This time, the local authorities were to take over development land and to keep some of the resulting profit, sharing the remainder with other local authorities and with the Treasury. But predictably, in 1979 the Conservatives came back and repealed the measure. However, since 1971 an alternative way of capturing gains in land value has developed: under section 106 of the Town and Country Planning Act 1990 so-called planning obligations (more popularly, planning agreements), whereby a developer agrees with the local planning authority to make a contribution, in money or kind, in return for grant of planning permission, are incurred.

Despite these many changes in the financial provisions of the 1947 Act, the main body of legislation has survived. (In fact all the 1947 provisions, together with subsequent amendments, were rolled up into consolidating Acts in 1962, 1971 and 1990, so that the '1947 Act' as such does not exist.) However, in Acts of 1968, 1991, 2004 and 2008, major changes were made in the way plans are prepared. We shall discuss those changes in the appropriate place, in Chapter 6.

The 1949 National Parks and Access to the Countryside Act

One further piece of legislation remained to complete the structure. The 1947 Act had at last given to local authorities strong powers to regulate the use of land in the countryside, and thus preserve fine landscape for the enjoyment of the community. But more positive action was thought to be needed on at least two fronts. First, certain especially fine areas needed to be planned in a special way for the enjoyment and recreation of the nation; and second, provision was needed to open up the countryside generally to the public, since (especially on many upland areas) they found themselves barred by sporting or other private interests.

Both the Dower and Hobhouse Reports, as we saw, assumed that national parks should be set up on the model already existing in America, with a strong national executive agency well provided with national funds to make large-scale investment for tourism and outdoor recreation – hostels, camp sites, trails and so on – and, most importantly, with the power to acquire land: the Hobhouse Report had assumed that about one tenth of the area of the parks should be acquired within ten years. But when the government established the parks, in the National Parks and Access to the Countryside Act of 1949, it fought shy of this radical step. In the 1947 Act it had just established local planning authorities based on the counties and county boroughs, it argued, and these were the appropriate bodies to plan the parks. In the case of parks overrunning county boundaries, provision could be made for joint boards. The only special arrangements made for the parks were two. First, a National Parks Commission was set up, to be financed from central government funds, with the responsibility of planning the general programme for the establishment and management of the parks; but in relation to the local authorities, its functions were merely advisory. It did have the power to channel subsidies to these authorities for certain defined purposes of development in the parks, such as car parks or information centres, but only when local funds were forthcoming to match the grants. And second, the Act provided for minority membership of outside interests – recreationists, conservationists and amenity organizations – on the local planning committees responsible for the parks. But these committees, in turn, reported to their full councils, which did not contain the outsiders.

Critics at the time attacked the proposals for being weak and insufficiently positive: the local authorities, they argued, were not likely in most cases to support a positive programme of developing the parks for national use, especially when this involved a burden on local rates. This fear proved only too well grounded. In the two decades after the Act, expenditure was negligible, and most of it was concentrated in the two parks which happened to be managed by joint boards: the Peak District and the Lake District.

The 1949 Act also gave local planning authorities in general some additional powers and responsibilities. They were to negotiate with landowners for the development of long-distance footpaths across areas of fine scenery, such as the Pennines or the coasts of Cornwall; the National Parks Commission, again, was to take the lead in developing a plan for their establishment. The plan was quickly forthcoming, but again progress was very slow: in the 23 years to 1972 only five such paths were actually established and opened. Planning authorities were also enjoined to designate areas of outstanding natural beauty, which were areas not justifying the full national park treatment, but nevertheless requiring a very special degree of strict planning control to prevent obtrusive or alien development. Since this was the sort of negative control the local planning authorities were well capable of exercising under the 1947 Act, not requiring agreement of landowners or the expenditure of local funds, it proved to be one of the more successful provisions of the Act.

One further development at this time was important for the planning of the countryside. In 1948 the government responded to the promptings of scientists and conservationists, and set up the Nature Conservancy by royal charter. It was given the power, and a fairly generous budget, to set up national nature reserves for the conservation of natural habitats and of wildlife, either buying the land for the purpose or reaching agreements with the landowner which would preserve the land in its natural state. On these reserves, and elsewhere, it developed an ambitious research programme, both with its own scientists and through university contracts. In 1965 the Conservancy became one arm of the newly formed Natural Environment Research Council, charged with the coordination of all research in that field. The marked success of the Conservancy, compared with the relative weakness of the National Parks Commission, demonstrated to many observers the importance of creating a strong executive agency with adequate funding and the power to spend it – though in 1972–3 the research and management functions of the Conservancy were split.

A tentative verdict

A considered verdict on the '1947 system' – as it is convenient to call it, after its central piece of legislation – clearly has to wait until after a study of how it worked in practice, which we shall do in Chapters 5 and 6. But meanwhile it is helpful to point out a few important features.

First, the system worked by giving strong negative powers of control to the new local planning authorities. Good positive planning, it seems to have been assumed, would mainly be carried out by public building agencies of various kinds – the local authorities and the new towns – in which close and virtually automatic union of planning and development would be the rule. They would be almost wholly responsible for the urban renewal programmes in the older parts of the cities and for the construction of new and expanded towns of all kinds in the countryside. The negative powers of control would be needed merely to control the minority of developments that would still be carried out by private agencies. In practice, as we shall see in Chapter 6, it worked out very differently.

Second, the system clearly required some overall coordination. It was generally agreed that the right unit for spatial or physical planning was the urban region, as Geddes had suggested as long ago as 1915. There was not much empirical work at that time on the delimitation of the spheres of influence of cities; the first serious empirical work, by A.E. Smailes and F.H.W. Green, was published round about the time the system was being set up. But clearly it extended right outside the rather restrictive boundaries of the cities, into the surrounding countryside; and Abercrombie's plan for Greater London covered an area of over 2,000 square miles (5,000 square kilometres). Nevertheless, in the 1947 Act the government gave the local planning powers to the existing local authorities, not to bigger units. This was probably inevitable if planning was to be accountable to a local electorate; but in addition, despite considerable interest in the idea, the government attempted no fundamental reform of local government. The critical job of preparing plans for the orderly development of the great city regions – and in particular for the decentralization of population and jobs in them – was split between the county boroughs and the rural counties. Though there was provision for joint planning boards, these were not implemented save on an advisory basis, and although by 1948 regional plans existed for the areas around all the great conurbations, prepared either by outside consultants or by joint committees of the local authorities concerned, they were purely advisory in function.

In these circumstances some coordination from the top was clearly essential. The 1947 Act provided it, by the requirement that plans be submitted to the minister for approval; the minister could amend them as s/he wished. Thus the various plans for any region could be coordinated. But for this to be done effectively, some kind of regional intelligence agency was clearly necessary to provide the minister with advice. The original organization of the Ministry of Town and Country Planning, with strong regional offices in each of the main provincial cities, was specifically created to deal with this. But during the 1950s these offices were closed down for reasons of economy, and at this point the idea of coordinating the various local plans seems to have been more or less abandoned. The almost inevitable result was that the various local planning authorities, left to their own devices, pursued a defensive and negative policy. We shall trace some of the consequences in Chapter 6.

Further reading

The best source for Barlow is still the report itself: *Report of the Royal Commission on the Distribution of the Industrial Population* (Cmd 6153, HMSO 1940; reprinted 1960). See also G.C. Allen, *British Industries and their Organization* (Longman, fifth edition, 1970). On land use, see L. Dudley Stamp, *The Land of Britain: Its Use and Misuse* (Longman, second edition, 1962), Chapter 21. On the beginnings of regional policy see Christopher M. Law, *British Regional Development Since World War I* (Methuen, 1981).

J.B. Cullingworth and Vincent Nadin, *Town and Country Planning in England and Wales* (Routledge, 13th edition, 2000), is the standard text on the British planning system. Gordon Cherry, *Town Planning in Britain Since 1900* (Blackwell, 1996) is an indispensable general source. The legislative base is set out in detail in Desmond Heap, *An Outline of Planning Law* (Sweet & Maxwell, 1959–, continuously updated). On new towns, see Frederic J. Osborn and Arnold Whittick, *The New Towns: the Answer to Megalopolis* (Leonard Hill, second edition, 1969); Frank Schaffer, *The New Town Story* (MacGibbon & Kee, 1970); Pierre Merlin, *New Towns* (Methuen, 1971), which also deals with experiments in other countries; Meryl Aldridge, *The British New Towns: A Programme without a Policy* (Routledge, 1979); Denis Hardy, *From Garden Cities to New Towns: Campaigning for Town and Country Planning, 1899–1946* (Spon, 1991) and Peter Hall and Colin Ward, *Sociable Cities: The Legacy of Ebenezer Howard* (Wiley, 1998). On urban containment, see David Thomas, *London's Green Belt* (Faber, 1970) and Peter Hall, Ray Thomas, Harry Gracey and Roy Drewett, *The Containment of Urban England* (Allen & Unwin, 1973).

5 National/regional planning from 1945 to 2010

In Chapter 4 we saw that the Distribution of Industry Act of 1945 effectively carried out the recommendations of the Barlow minority report: there was to be a strong policy of steering industrial growth from the more prosperous regions to the depressed areas of the 1930s, to be accomplished not only by positive inducements to locate in these latter areas, but also by negative controls over the location of new industry, and over extensions to existing industry, in other areas. This control applied only to manufacturing industry, though more rapid growth was in the tertiary or service sector of employment; and this was not at all remedied until the control of office development in 1964.

In fact, little substantial change was made in the control mechanisms from 1945 to 1960, except for adjustments in the lower thresholds of size below which no industrial development certificate (IDC) was needed. There is, however, clear evidence from the statistics of the Board of Trade that the whole policy of steering industry was operated rather more laxly in the 1950s than in the period from 1945 to 1950. This might be attributed to the fact that a Labour government was more enthusiastic about helping the development areas (where much of its voting support was concentrated) than a Conservative government; the more likely reason is simply that by the early 1950s it seemed clear that general economic management policies were keeping unemployment levels well below the levels of the 1930s, so that the case for strong regional policies seemed rather weaker than in 1945.

In this chapter, therefore, we will look first at the record of the controls, and their effects, from 1945 to about 1980. Then we shall turn to look in some detail at the rapid – and sometimes bewildering – policy shifts of the 1960s and 1970s, and try in turn to sum up their effects. We then turn to the radical reversal of previous policies in the 1980s: a reversal which in effect abandoned the regional policies in force since 1945. Finally, we consider some of the policy shifts evident since 1997, including the development of a stronger regional policy and regional governance framework, and the way that large national infrastructure projects are determined.

Regional policy and regional change, 1945–80

An elementary point should first be made: that almost throughout the period from 1945 to 1980, the overwhelmingly most important aim of planning policy at the national/regional scale was to create employment. More precisely, it was to reduce unemployment rates and/or the rates of out-migration from the development areas. There are a number of possible aims of regional economic policy: they include improving the efficiency of industry, raising the level of gross regional product per worker or per head of total population, improving the distribution of regional income and many other variants. Trying to keep employment up (or unemployment down) could in fact easily run counter to many of these other objectives. It could, for instance, lead to the retention or even the introduction of rather inefficient labour-intensive industries that paid poorly, thus keeping a large section of the population in low-income occupation and increasing the inequality of income within the region, as well as the inequality between that region and the rest of the country. Many economists would argue, indeed, that the obsession with employment as almost the sole criterion of British regional policy has been positively pernicious.

However, there are two obvious reasons why this objective has been so attractive. The first is that unemployment is much more visible than low income or inequality in income. People are less inclined to put up with it, and politicians are therefore more concerned about it. The second is that at least for much of this period the statistics with which to measure other criteria of regional performance were poor or non-existent. This particularly applies to figures about regional productivity, which were few in number and late to appear. In any case, calculations of productivity, unless they are accompanied by very full statistical information about some of the possible explanations – such as the amount and quality of capital and the training of the labour force – are notoriously dangerous to interpret.

We shall, therefore, concentrate on the employment criterion, as contemporaries did, with a sideways look at other possible indices. Table 5.1 shows the actual results of the Board of Trade's operation of the IDC machinery from 1956 to 1960 and from 1966–70. (Comparable figures are unfortunately not available for the 1970s). It shows fairly clearly that under Conservative governments the 1950s was a period of weak or hands-off regional policy: the machinery remained in existence but it was not operated very actively on behalf of the development areas. But in the 1960s, a period of Labour government, there was a systematic diversion of new factory floorspace from the prosperous areas to the less prosperous areas. A measure of the amount of this diversion can be obtained by comparing the proportion of new floorspace in each region with the yardstick of the employment in that region at the beginning of the period. Thus the East and West Midlands, the South West and the south-east corner of England, consisting of the South East and East Anglia regions, had together 58.8 per cent of total national employment at the start of the 1966–70 period; but they obtained only 41.5 per cent of new floorspace in the following four years. Conversely, the North region obtained no less than 10.9 per cent against the yardstick of 5.7 per cent; Scotland 14.0 per cent against 9.3 per cent; and Wales 8.6 per cent against 4.3 per cent. Dividing the whole country up into the 'more prosperous' regions of the South and Midlands, and the 'less prosperous' regions of the North, Scotland and Wales (that is, the areas where the development areas were concentrated), the distinction is clear: the less prosperous regions got 58.8 per cent of new floorspace against a yardstick of 41.2 per cent. (In the 1956–60 period, contrastingly, they got only 42.7 per cent against a yardstick of 43.2 per cent.) We can assume that the distribution of new factory jobs followed the distribution of new factory floorspace – though not precisely, since some at least of the new factory space was in capital-intensive, labour-saving types of production.

But the picture looks rather different when we turn to the right-hand column of Table 5.1. This shows the actual creation of employment in the regions – all employment, not just factory jobs. In the 1950s, when the south-east corner of England received only 30.9 per cent of new floorspace, it attracted 58.2 per cent of all new jobs. Between them, five less prosperous regions – the North West, Yorkshire, the North, Scotland and Wales – together got 42.7 per cent of new floorspace, but only 11.5 per cent of the new jobs: a derisory total. In the later 1960s, the comparison is complicated by the fact that employment was almost everywhere falling. But the same conclusions apply; together the five less prosperous regions had 58.8 per cent of additional floorspace, yet more than half of the total loss in jobs was concentrated here.

Table 5.1 *Industrial building completions and employment changes, 1956–60 and 1966–70*

Old standard region[a]	Employment at start of period No. (thousands)	% total	1956–60 Industrial building completions million sq. ft	% total	Employment changes No. (thousands)	% total gain
Northern	1,279	5.9	13.2	5.6	+23	3.7
East & West Riding	1,857	8.5	18.9	8.0	+17	2.7
North West	2,983	13.7	32.4	13.7	+10	1.6
Wales	956	4.4	14.8	6.2	+13	2.1
Scotland	2,163	10.0	21.7	9.2	+9	1.4
'Peripheral' regions	9,238	42.6	101.0	42.7	+72	11.5
London & South East, Eastern & Southern	7,633	35.2	73.1	30.9	+365	58.2
South West	1,189	5.5	11.6	4.9	+57	9.1
West Midlands	2,148	9.9	33.5	14.1	+83	13.2
North Midlands	1,485	6.8	17.9	7.6	+53	8.5
'Prosperous' regions	12,455	57.4	136.1	57.5	+558	89.0
Great Britain	21,706	100.0	236.9	100.0	+627	100.0

New standard region[a]	Employment at start of period No. (thousands)	% total	1966–70 Industrial building completions million sq. ft	% total	Employment changes No. (thousands)	% total gain
Northern	1,335	5.7	20.7	10.9	−14	2.2
Yorkshire & Humberside	2,111	9.0	19.3	10.2	−87	13.9
North West	3,034	12.9	28.6	15.1	−129	20.6
Wales	1,007	4.3	16.3	8.6	−45	7.2
Scotland	2,193	9.3	26.6	14.0	−41	6.5
'Peripheral' regions	9,680	41.2	111.5	58.8	−316	50.4
South East	8,068	34.3	30.1	15.9	−212	33.9
East Anglia	615	2.6	7.9	4.2	+29	−4.6
South West	1,355	5.7	10.7	5.6	−22	3.5
West Midlands	2,388	10.1	16.6	8.8	−95	15.2
East Midlands	1,437	6.1	13.3	7.0	−8	1.3
'Prosperous' regions	13,863	58.8	78.6	41.5	−308	49.3
Great Britain	23,554	100.0	189.5	100.0	−626	100.0

[a] The 'old' map of the Objective One regions, as explained above, were abolished in 1965 – a major statistical problem in comparison.

Source: *Abstracts of Regional Statistics*

Note: Total may not add up, owing to rounding

The simple reason, of course, is that most of the new jobs were not factory jobs. They were in services. Table 5.2 is a 'league table' of the 24 main orders of the Standard Industrial Classification in the period 1960–2001, ordered in terms of the percentage increase in employment. During much of this period, the two fastest-growing employment groups were in the tertiary or service sector: professional and scientific services, and insurance, banking and finance. And these were not cases of big increases on small bases: both were among the more important industrial groups in the national economy. By and large, these jobs – together with some of the fastest-growing factory jobs – were particularly well represented in the more prosperous regions. Conversely, the development areas tended to have higher proportions of people in the stagnant or declining industries which occur at the foot of the league in Table 5.2. These tend to be the same older staple industries, like coal mining and textiles and shipbuilding, whose decline caused such acute distress in these areas in the 1930s. The problem of bad industrial structure, it seems, had not been eradicated.

This conclusion poses the question, first raised by Barlow: was the poorer performance of the problem regions, in terms of employment growth, wholly to be explained by this effect of industrial structure? Even a casual glance at detailed tables of regional industrial structure suggests that it is more complex than that: a region like the West Midlands has much-better-than-average growth, though apparently it has a lower-than-average representation of growth industry; conversely a region like Scotland has a poorer-than-average performance, though it has a better-than-average proportion of fast-growing industry. Several analyses by economists during the 1960s tried to calculate the importance of the structural effect, and have concluded that though in some regions it seemed to have dominated – especially in the poor performance of the problem regions – elsewhere it was quite unimportant. Figure 5.1 shows one such set of results, published by Frank Stilwell in 1969. But this type of statistical analysis – it is called 'shift-share' analysis – is quite abnormally sensitive to the classification of industry that is used. If the grouping of industry is very coarse, with a number of rather disparate industries having different location patterns lumped into one group, the structural effect is much less likely to show up than if the classification is a fine one; and the one used in the analysis described here was quite a coarse one.

The sluggish growth in employment in the less prosperous regions would not matter so much, of course, if it were in line with the demand for jobs. But there is clear evidence that it is not. In the first place, all these regions have continued to experience higher rates of unemployment, on average, than the rest of the country. The differential is much smaller than in the days of very high national unemployment rates during the 1930s; but it is there, and what is significant is that whenever the national rate widens, then the differential of the less prosperous areas widens too (Figure 5.2). What is also evident from Table 5.3 is that the unemployment rate does not measure the full extent of the true waste of labour. For in addition, activity or participation rates also tend to be lower in some of these regions, especially for women. These rates simply measure the proportion of adult men and women actually in the labour force (whether employed or unemployed). They show the extent to which the region is tapping its reserves of labour. Of course, female activity rates, in particular, can vary because of social customs: in some mining areas it has traditionally been thought that the woman belonged at home, not in the factory. But to a large extent, especially in recent years, they tend to reflect simply the availability of work: where the rate is low, it means that some people think there is no point looking for a job.

In many ways, the regional income per head is the best of all indices of economic health or the lack of it. It sums up many different causes working in conjunction: high unemployment, low activity rates, poor industrial structures with large proportions of

Table 5.2 *Total employment changes by industrial orders: United Kingdom, 1960–70, 1971–8, 1991–2001*

Old industrial order	1960–70		1971–8 (decennial rate)	
	Per cent	Order	Per cent	Order
Agriculture, forestry, fishing	−41.4	26	−14.9	18
Mining and quarrying	−46.1	27	−19.1	23
TOTAL: PRIMARY INDUSTRIES	−44.0		−11.8	
Food, drink, and tobacco	+7.6	10	−9.7	14
Coal, petroleum products	−2.7	16	−26.0	26
Chemicals and allied industries	+1.9	13	−2.7	9
Metal manufacture	−4.8	18	−25.1	20
Mechanical engineering	+7.3	11	−15.7	20
Instrument engineering	+9.2	8	−15.4	19
Electrical engineering	+15.6	6	−10.9	16
Shipbuilding and marine engineering	−31.8	25	−6.7	11
Vehicles	−8.9	20	−7.9	12
Other metal goods	+16.2	5	−8.9	13
Textiles	−21.9	3	−29.2	27
Leather, leather goods and fur	−15.6	23	−18.3	22
Clothing and footwear	−16.1	24	−22.0	24
Bricks, pottery, glass, cement, etc.	−0.9	15	−17.7	21
Timber, furniture, etc.	+1.0	14	−2.6	10
Paper, printing and publishing	+8.1	9	−12.7	17
Other manufacturing industries	+17.3	4	−1.3	8
TOTAL: MANUFACTURING INDUSTRIES	−0.4		−9.4	
TOTAL: CONSTRUCTION	−9.2	21	+1.0	7
Gas, electricity and water	+2.6	12	+10.3	5
Transport and communication	−4.6	17	−10.4	15
Distributive trades	−5.7	19	+7.0	6
Insurance, banking and finance	+75.3	1	+25.7	3
Professional and scientific services	+46.6	2	+32.9	2
Miscellaneous services	−9.6	22	+34.7	1
Public administration	+10.4	7	+12.1	4
TOTAL: SERVICE INDUSTRIES	+6.9		+11.0	
• Excluding adult students				
GRAND TOTAL	+0.8		+2.7	

New industrial order	1991–2001	
	Per cent	Order
Agriculture, forestry, fishing	−1.8	6
Energy and water supply	−54.5	9
TOTAL: PRIMARY INDUSTRIES	14.4	
TOTAL: MANUFACTURING INDUSTRIES	−15.0	8
Construction	−2.8	7
Distribution, hotels and catering, repairs	14.3	3
Transport and communication	8.8	4
Banking, finance, insurance, business services	27.0	1
Public administration	6.8	5
Other services	26.1	2
TOTAL: SERVICE INDUSTRIES	14.8	
GRAND TOTAL	7.0	

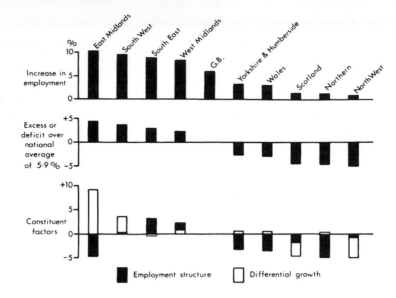

Figure 5.1 The structural effect on regional employment change, 1957–67. Broadly, the regions of southern and midland England have had faster than average employment growth while the North, Wales and Scotland have lagged. Calculations by the economist Frank Stilwell show that in several cases – especially in the laggard regions – their poorer performance can be explained largely in terms of the unfavourable economic structure, with a predominance of declining or static industry.

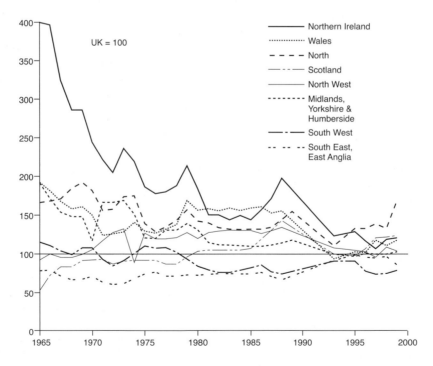

Figure 5.2 Regional unemployment differentials, 1961–2000. Generally, since 1961 differentials have narrowed, but there have been many perturbations and the less prosperous regions – the North, Wales, Scotland and Northern Ireland – are still less noticeably worse off than the South and Midlands.

Table 5.3 *Regional relatives for unemployment, activity rates and incomes, 1999*

	Unemployment	Female activity rate	Household income
United Kingdom	100.0	100.0	100.0
North East	168.3	92.9	85.9
North West	103.3	96.0	92.5
Yorkshire and the Humber	108.3	99.0	92.2
East Midlands	86.6	103.1	94.9
West Midlands	113.3	100.8	91.9
East	68.3	103.3	110.3
London	126.6	97.1	117.9
South East	60.0	105.5	112.3
South West	78.3	106.2	101.5
Wales	116.6	93.9	87.4
Scotland	123.3	99.2	92.1
Northern Ireland	119.9	89.9	90.0

low-paying jobs. Incomes can be presented in a number of different ways, from different sources. Table 5.3 shows two of them, and it is fairly clear from it that high unemployment and low female activity rates are associated rather systematically with low household incomes. Figure 5.3 shows that in general the South and Midlands have higher incomes, while the problem areas still have lower average incomes. Worst off of all, it seems, are the thinly populated upland rural areas such as mid-Wales, which have large numbers of small-scale hill farmers subsisting on very low incomes.

Faced with the prospect of higher-than-average unemployment risk, fewer job opportunities for women and lower incomes, it is small wonder that many people choose to leave the problem areas. But here again a word of caution should be entered. The net figures of inter-regional migration show very clearly that the broad drift is out of the problem areas and into the more prosperous South and Midlands. But these net figures are in fact relatively small differences between much larger gross flows (Table 5.4). The North East, for instance, lost 44,000 people in 1998. But in the same year it gained 39,000 people into the region; thus, there was a net outflow of 5,000. What the table does not show is immigration from abroad. London, for instance, suffered a net outflow of 47,000 people in 1998. But, as in every year in the 1990s, there was an approximately equal net inflow of people from abroad – and, since London's young population gave it a high birth rate, its population was rising again for the first time in 50 years.

The conclusion that could be drawn concerning regional economic policy by the middle 1970s, then, was that it had worked hard to provide new jobs in the problem areas, but that it had made relatively little difference to the overall picture. Of course, it should be stressed that if the policy had not operated, matters would doubtless have been that much the worse. A definitive study, published in 1973 by Barry Rhodes and John Moore, compared the employment created in the assisted areas in two different periods: one of weak regional policy between 1951 and 1963, the other of strong policy between 1963 and 1970. It concluded that between 1963 and 1970 regional policies had created some 250,000–300,000 jobs in the assisted areas that would have been created otherwise. Subsequent work by Brian Ashcroft and Jim Taylor has suggested that the actual rate of job creation was less than this, perhaps as little as 90,000. Against this, of course, there was a considerable monetary outlay by government: regional development

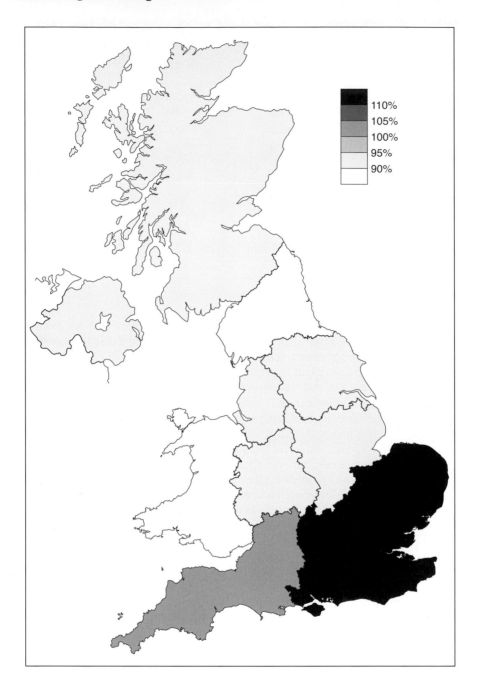

110%
105%
100%
95%
90%

Figure 5.3 Income per head index, 1997. Maps of regional income disparities reinforce the view that 'two nations' still exist in Britain. The most prosperous part is 'Megalopolis', stretching from London through the Midlands to Cheshire; the peripheral regions (except oil-rich Aberdeen) are noticeably poorer.

Table 5.4 *Migration, gross and net, between standard regions, 1998 (thousands)*

	Gross immigrants	Gross emigrants	Net migration
North East	39	44	–5
North West	104	116	–12
Yorkshire and the Humber	93	98	–5
East Midlands	108	97	+11
West Midlands	93	101	–8
East	143	124	+19
London	171	218	–47
South East	226	207	+19
South West	139	111	+28
Wales	56	54	+2
Scotland	53	54	–1
Northern Ireland	12	12	0

Table 5.5 *Regional selective assistance to manufacturing, 1989/90–1998/9, £ million*

	1989/ 90	1990/ 91	1991/ 92	1992/ 93	1993/ 94	1994/ 95	1995/ 96	1996/ 97	1997/ 98	1998/ 99
North East	117.0	85.0	63.8	48.3	52.7	38.4	46.4	24.3	38.1	22.3
North West	74.3	57.5	49.5	36.8	40.3	32.4	24.3	23.2	19.4	25.9
Yorkshire and the Humber	32.4	29.4	18.2	13.7	35.6	23.0	19.7	11.1	12.7	11.9
East Midlands	9.5	5.5	2.6	1.2	1.9	5.2	7.3	10.5	10.5	7.1
West Midlands	19.9	18.0	8.7	10.8	14.4	14.7	14.2	25.5	29.8	30.6
East	—	—	—	—	—	0.7	2.1	1.5	2.2	0.7
London	—	—	—	—	—	0.6	1.7	2.9	2.7	3.2
South East	—	—	—	—	—	0.9	4.2	4.1	5.4	3.3
South West	10.7	9.0	8.3	8.2	9.5	9.4	7.7	7.4	4.5	9.4
England	263.8	204.4	151.1	119.0	154.4	125.3	127.6	110.5	125.3	114.4
Wales	131.7	133.7	153.9	140.6	118.8	109.2	98.0	132.4	172.6	153.9
Scotland	143.8	159.2	122.8	104.4	121.2	134.4	117.4	128.2	132.5	125.5
Northern Ireland	127.1	132.1	138.0	105.6	117.6	132.9	131.2	137.1	156.1	153.3
Great Britain	539.3	497.3	427.8	364.0	394.4	368.9	343.0	371.1	430.4	393.8

grants rose to around £400 million a year by the late 1970s, and were running at much the same rate throughout the 1990s (Table 5.5). But such outlays are of course transfer costs: Moore and Rhodes argue that the true resource costs of the policies were close to zero.

At the same time, most new jobs were still being created outside the development areas. For this there were two reasons: growth in service-industry employment not subject to the IDC controls, and growth in factory industry which in one way or another fell outside the IDC net. A.E. Holmans, for instance, calculated in 1964 that of an increase of 577,000 factory jobs in south-east England (including East Anglia) in the 1950s, only 190,000 could be accounted for by the grant of IDCs; the rest had been created through small-scale extensions that escaped the controls, or through buying up existing factory buildings. In addition, Holmans pointed out, this area had the great bulk of all employment in two of the fastest-growing service industries, professional and scientific services and miscellaneous services: with 65 per cent of total national employment in these groups in 1959, it had 65 per cent of the subsequent growth from then to 1963. Though the rate of growth of these groups was no higher in the South East than elsewhere, they contributed importantly to its favourable overall growth record, and though the government made a partial response in 1965 by trying to control other employment, there was little evidence that this had much direct impact on regional job creation.

Policy changes, 1960–80

During one part of the period just analysed, up to 1960, there was considerable stability in the regional economic policy framework – though, as already seen, there were big shifts in the way in which it was used. Both the policies and the regions to which they related remained constant except in minor detail. The distinction between the development areas and the rest was quite fundamental, and it was maintained as the 1945 Distribution of Industry Act had determined. These development areas – Merseyside, the North East, West Cumberland, central Scotland and Dundee, and South Wales – had been fixed in the 1945 Act; they covered quite broad areas of territory, roughly the older heavy industrial areas based on the coalfields, which had been designated as special areas in 1934 (Figure 5.4; Plates 5.1, 5.2 and 5.3).

The first major policy shift, contained in the Local Employment Act of 1960, changed all this (Figure 5.4b). It scrapped the development areas, and replaced them by the development districts, a more flexible concept defined as an area which had suffered a 4.5 per cent (or worse) unemployment rate over a sustained period of several months. The intention, laudable in principle, was to concentrate help on the distressed areas where it was most needed, and conversely to avoid helping those parts of the old development areas – especially the bigger commercial centres, which were seats of expanding service industries – that could well help themselves. But in practice the policy had all sorts of unfortunate effects. It proved to spread help very widely and thinly over sparsely populated rural areas, such as the Highlands of Scotland. This in itself might have been justified, because presumably such areas were in need of help. But by being applied rigidly to each and every local employment-exchange area which qualified on the unemployment criterion, it hindered the development of a concentrated strategy. Worse, by excluding the more prosperous local centres it frequently made it difficult to devise any strategy based on the natural point of growth in a region.

It was in fact just at this time – the early 1960s – that British economic planners and economic geographers began to show interest in the concept of the 'growth pole' (*pôle de croissance*) which had been developed in 1955 by the French economist François Perroux. They almost certainly mistranslated and misunderstood the concept: the 'pole' of Perroux was the sector of an economy rather than a geographical place or area, but in Britain it was interpreted in the latter sense. The result, naturally, was widespread professional criticism of the operation of the 1960 Act, on the ground that it inhibited the development of a 'growth pole' policy for the less prosperous regions. And, to make matters urgent, the winter of 1962–3 marked a serious recession in the national economy, which caused unemployment rates in these regions to swing up sharply.

The outcome, in November 1963, was the publication of government plans for two of the most seriously hit regions: the North East and central Scotland. Both plans, in

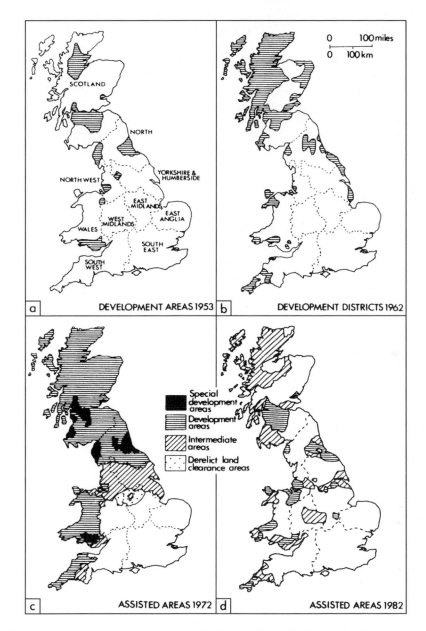

Figure 5.4 The pattern of regional development in Britain, 1945–88: (a) the original postwar scheme of closely defined development areas was replaced in 1960 by (b) development districts based on a criterion of persistent unemployment, and then in 1966 by (c) more generously defined development areas; later, special development areas were designated within these, qualifying for more generous state aid. But (d) after 1979 the assisted areas were sharply reduced in size, so as to concentrate on the worst-afflicted places.

Plate 5.1 Aerial view of Port Talbot steel mill, Glamorgan, South Wales, c. 1970 – an example of the new industry deliberately implanted in a development area. This picture also illustrates the attraction of deep water for heavy industry in postwar Britain.

Plate 5.2 The former Ford factory at Halewood, Merseyside (later to become a Jaguar manufacturing site). The then modern plant outside Liverpool was located there in the 1960s, with government encouragement, to implant one of Britain's growth industries in a development area. Similar development took place on the opposite bank of the Mersey at Ellesmere Port (General Motors), Linwood in Scotland (Chrysler) and Bathgate in Scotland (British Leyland).

Plate 5.3 Reconstruction in central Glasgow. To clear the backlog of obsolescence in Scotland's biggest city was a Herculean task. Government policies in the 1960s and 1970s deliberately diverted funds into the city's ambitious urban motorway programme, seen here; the objective was to give a new image to the city and to Scotland, but, in turn, it created a rather different form of environmental degradation and community impacts.

effect, represented partial abandonment of the 1960 development district policy, and its replacement by the concept of the growth pole – or, as it was translated, growth zone. The idea was to identify those parts of the region which had the best prospects of rapid industrial growth and to concentrate help – especially public investments in infrastructure (transport, communications, power lines and the like) – on them. The other areas, in effect, would be treated as virtually beyond help. But since they would normally be quite near the growth zones, their populations could quite easily readjust over time either by short-distance migration or by commuting to work there. In effect this neatly reversed the 1960 policy. For instance, in the North East the 1960 Act had concentrated help on the struggling western industrial districts of County Durham, west of the Great North Road; the 1963 Plan concentrated most of the help east of it.

At the same time, there was widespread criticism of the continuing disparity between the distress in the problem areas and the continuing rapid growth of the more prosperous areas – above all London and the South East. The 1961 census revealed that the area embracing London and the ring 50 miles around had added 500,000 to its population in the 1950s. Much of this, in fact, represented either natural increase or immigration from the Commonwealth; but that was not generally noticed, and it was widely assumed that London's prosperity was somehow connected with the North's distress. Particular criticism was directed at the office building boom in London, which was adding 3.5 million square feet of floorspace a year in the late 1950s, and which was certainly instrumental in the continuing increase of employment in the South East.

Prodded, the government began to take action. It sealed an absurd loophole in the planning regulations, which had allowed office developers to put back much more on a site than they demolished. Then in 1963 it set up the Location of Offices Bureau (LOB) to act as an information and publicity centre to encourage office firms to move voluntarily out of London. LOB's vigorous and compelling propaganda campaign, which seldom missed a topical opportunity, brought results: 10,000 jobs were being exported from the capital in this way each year during the late 1960s, most of them, however, to locations within 40 or 50 miles. And there was nothing to stop other users taking the offices they left.

The *South East Study* (HMSO, March 1964) added to the voices of the critics. An official study based on three years' work, it revealed that between 1961 and 1981 the region within 50 miles of London was likely to add 3.5 million to its population, and suggested a further serious of new towns to cope with the growth. The Labour Party, in particular, were highly critical of the situation, and on return to office in October 1964 they immediately began to review the study. Meanwhile, as of midnight on 4 November 1964, they imposed a ban on further office building in London and the region around it. In 1965 they formalized this emergency control in the Control of Office Employment Act, which at last put the creation of new office space on the same basis as the building of new factory floorspace. This gave them power to regulate new office building by the requirement that prospective builders apply for an office development permit (ODP). The controls applied initially to the London and Birmingham areas, and they were operated strictly. However, no powers at all were taken to cover any of the other fast-expanding sectors of tertiary industry, such as retailing or higher education. Much of the latter, ironically, was growing very rapidly in the South East, including the centre of London, as part of publicly approved and financed programmes.

During 1965, the new Labour government created a totally new organization for promoting and coordinating national/regional planning, as they had promised to do before the 1964 election. This, however, was less a party political platform (though the Labour Party was historically committed to helping the less prosperous areas) than the reflection of a general movement of thought among economists and planners. There was at this time an intense interest in the French system of economic planning, which had been in operation since 1946, and which had come to contain a very strong regional element. The French system was well adapted to a mixed economy with both public and private sectors, such as that of France itself or Britain. Called indicative planning (as distinct from regulatory planning), it relied heavily on the coordination of public and private investment programmes through a complex structure of councils and committees. (It is explained in more detail in Chapter 7.) In the system as it existed in the early 1960s, when British observers were studying it intensively, there were basically two sorts of coordination: by industrial group, and by region. For the latter purpose, the country was divided up into 21 planning regions, consisting of groups of *départements* (the basic system of French government in the provinces, though not 'local government' in the British sense), each under a director, or *super préfet*.

The 1965 reform was in essence an attempt to apply this system to Britain. Immediately on entering office the Labour government had set up a Department of Economic Affairs, which set to work on the preparation of a national plan of an indicative type. To provide the necessary element of regional coordination, they set up a series of economic planning councils, consisting of members appointed by the government to represent different groups in each region (industrialists, trades unionists, traders, transport workers, academics), and economic planning boards of civil servants seconded from their London departments to work together in the regions; each of the regions had a council charged with the preparation of a regional study and plan, assisted by the professional board members. The regions used for the purpose in England were basically the old standard regions which had been used for statistical purposes ever since the Second World War, with some detailed modifications in the South East and on Humberside. Wales and Scotland each constituted a region by itself.

In practice the record of the councils and boards was a mixed and not always a happy one: one council chairman resigned after disagreement with the government, at least one other threatened to do so, and many expressed private frustration over the relative powerlessness of the councils. The original National Plan was published before it could contain any contribution from the councils and boards. Subsequently every council published a regional study and most published a plan for economic development, but in many cases the government rejected their recommendations. More seriously, the Department of Economic Affairs (DEA) – which coordinated the work of the councils – became weaker after the departure of George Brown as its political head: and in 1969 it was formally abolished, its long-term economic planning functions passing to the Treasury and its regional responsibilities passing to the Ministry of Housing and Local Government.

In part, this can be attributed to inter-ministerial warfare: from the start, the Treasury disliked the idea of a rival economic department, and the Ministry of Housing was worried that physical planning would be subordinated to the new department. But more basically, the new structure had real difficulties. On the economic side it proved difficult to divide up economic planning – the short-term work staying with the Treasury, the longer-term plans going to the new DEA – as the government had thought possible in 1964. On the regional side it transpired that in many cases – especially in the more buoyant, faster-growing regions of the country – the work of the councils and boards had an extremely strong element of physical planning. After all, in a region like the South East, the main responsibility of economic planning must be to prepare a plan for the orderly decentralization of employment and population, which is taking place anyway; and this is a spatial or physical plan. One result was a demarcation dispute between the council and the existing physical planners, which is discussed in Chapter 6. The outcome was that by 1969 it seemed natural that the work of the councils and boards should be coordinated by the Ministry of Housing and Local Government – reflecting the fact that the main work in future would be in overall spatial planning of the regional/local variety, rather than as part of a national/regional planning exercise which seemed to have come to a sad end.

This story has taken us some way ahead of chronology, to which we should now return. Between 1966 and 1967 the Labour government made a major shift in the structure of incentives to firms moving to the problems areas. In the 1966 White Paper on Investment Incentives, and the resulting Industrial Development Act of 1966, they scrapped the development districts and replaced them again by development areas – though defined more widely than in the 1945–60 period. Indeed, they were defined so widely that they could be criticized on almost the same grounds as the development districts: that they offered the prospect of help to areas beyond help. However, they were an improvement in that (with one or two glaring exceptions) they did not exclude the growth centres in the regions from the possibility of aid. The new development areas took in the whole of northern England, north of a line from Morecambe Bay to Scarborough; all Scotland, save Edinburgh; all Wales, save the Cardiff and Newport areas and Flintshire; north Devon plus north Cornwall; and Merseyside. (Shortly after special development areas were defined for declining coal-mining regions such as Lancashire, West Cumberland and the Welsh valleys.) There was much detailed criticism of the boundaries, as was perhaps to be expected: it was pointed out that the exclusion of Cardiff made it difficult to prepare a rational regional plan for south Wales (and later – a more glaring anomaly – a new town was proposed at Llantrisant, north of Cardiff, on a site that was half inside the development area, half out); that Merseyside was really no worse off than the depressed cotton towns of north-east Lancashire, which got no aid at all; that it was wrong to treat the struggling coalfield towns of south Yorkshire on the same basis as the prosperous Home Counties; and so on. These criticisms caused the government to appoint a Committee of Inquiry on the Intermediate Areas – the areas like Lancashire and Yorkshire, which were more prosperous than most development areas but decidedly less prosperous than the South or Midlands, as many of the tables and diagrams earlier in this chapter show. We will consider its recommendations shortly.

Meanwhile, the 1966 Act stipulated that firms moving to the development areas, as now redefined, should receive a cash grant amounting to 40 per cent of the value of any investments in plant and machinery (as against 20 per cent elsewhere). This provision provoked some criticism on the ground that it would attract capital-intensive industry which would bring very little employment into the development areas; it was argued that the result could be to diminish employment, since the grant would be used to install automatic machinery. Responding to this charge, the government in 1967 proposed a radical new departure. This year previously they had introduced the Selective Employment Tax (SET) with the aim of diverting labour from service industry to manufacturing by means of a tax on all establishments in the tertiary sector. This tax originally did not have a regional component: it was nationally uniform, and its stated aim was to increase the productivity of labour, since economic experts argued that bigger productivity gains could be made in manufacturing (with efficient machinery and management) than in the often under-organized, under-capitalized service sector. But in 1967, the government proposed (and then introduced) an important modification: the Regional Employment Premium (REP). From then on, not only would manufacturing industry in the development areas not have to pay the tax; it would also receive money grants, amounting to nearly £100 per annum per worker in the case of male adults. This meant that the difference in status between development areas and non-development areas (and between manufacturing and service industries in the development areas) was very large; and it broadened further when in 1968, in a financial crisis, the government announced that all firms in non-development areas – even in manufacturing – would pay SET. At that point, in fact, SET–REP became a completely regional device to attract industry into the development areas – and a very powerful one.

This of course added to the complaints from the less prosperous areas outside the development areas – the grey areas, or as they were called in the official report on the problem in 1969, the intermediate areas. This report, commonly known as the Hunt Report after its chairman (Sir Joseph Hunt), recommended substantial additional help to Lancashire and Yorkshire, in the form of building grants and an end to all IDC restrictions in those areas; it also recommended withdrawing Merseyside's status as a development area, which it claimed was no longer justified. The government gave only limited help to the intermediate areas in its 1970 Act; and the considerable difference in status between the development areas (including Merseyside) and the rest persisted.

In 1970 the incoming Conservative government soon made it clear that it intended to scrap many of the policies of its predecessor with regard to regional incentives. It announced a phasing-out of SET and REP altogether from 1974 and therefore an end to any differential applied to labour costs. It also announced an end to investment grants, replacing them by tax allowances on new investment. It estimated that the end of REP would save some £100 million a year in government expenditure, despite loss of the revenue from SET; ending investment grants would save £400 million a year in 1972–3, offset by £200 million a year for the new allowances. Later, though, in the 1972 Industry Act – impelled by rapidly rising unemployment – it re-introduced investment grants (entitled regional development grants) in the development areas, special development areas and intermediate areas. The government made it clear that much of its industrial policy would be directed at the modernization of industrial plant everywhere in the country through investment allowances, free depreciation for plant and a high initial allowance for buildings – even where no notable increase in employment was in prospect. Creation of jobs in the development areas was no longer the overriding aim; even the regional development grants would not depend on this criterion.

The Labour government, from 1974 to 1979, continued to operate the Conservatives' 1972 Act, though far from dismantling the REP, they at first increased it. (Then, in 1976, after the UK had entered the European Economic Community in 1973, it had to be abolished as contrary to EEC rules.) The main shifts in regional policy represented a direct – and progressive – response to the fact of rising unemployment. The Temporary Employment Subsidy of 1975 aimed to postpone redundancies in the assisted areas: the Job Creation Programme, in the same year, aimed to stimulate the creation of new jobs. In the following year they were progressively extended in scope and given bigger funds (Figure 5.4c).

Meanwhile, however, the nature of the regional problem itself was changing. More and more, a new distinction was emerging: between the older conurbation cities, whether in the assisted areas or outside them, and the suburbs plus free-standing smaller cities. The inner cities were experiencing massive losses of people and jobs, much of the latter loss being due not to movements of plants, but to simple closure. The indiscriminate nature of regional aid, covering as it did more than half the country by the 1970s, did not at all recognize this distinction, so that in practice less and less of this aid went to the hardest-hit inner-city areas. We shall look in more detail at this problem in Chapter 6.

The great policy reversal of the 1980s

Soon after election to office in 1979, the new Conservative government embarked on a radical reconstruction of the map of regional aid. The boundaries of the assisted areas were cut back; previously covering some 43 per cent of the United Kingdom population, they now accounted for only 25 per cent (Figure 5.4d). They focused sharply on the hardest-hit industrial conurbations of Scotland and the North. Within these areas aid was to be maintained, so that the difference between the remaining assisted areas and the rest would henceforth be very sharp. Finally, however, the requirement to obtain an IDC was lifted for all developments of under 50,000 square feet; that is, from the great majority of all factory-building projects. Thus, while regional policy still offered a reduced carrot, there was virtually no stick whatsoever. All in all, the cost savings for the new policy were estimated to amount to more than one third of the £609 million budget provided for 1982–3.

In place of regional policy, the Thatcher governments offered urban policies that were highly targeted at the most problematic parts of the inner cities, which had suffered dramatic losses in employment as the result of structural economic changes in the second half of the 1970s and the first half of the 1980s: notably, the loss of manufacturing industry and of port and other goods-handling functions. But, since these policies represented a continuation of a trend established by the previous Labour government from 1977 onwards – albeit with very different policy measures – they are better discussed in Chapter 6.

Meanwhile, regional policy initiatives increasingly passed to the European Commission in Brussels. In the late 1970s, reflecting the changed balance of interests after the UK's accession to the European Economic Community (EEC) in 1973, the Common Agricultural Policy (CAP) was balanced and supplemented by structural funds designed to help declining regions make a successful shift to new and viable sources of economic growth. In practice, as will be seen in Chapter 7, the funds have been applied to assist both peripheral rural areas of low-income peasant farming (like Ireland, Spain and Portugal, Greece and, in the 1990s, north-east Germany, the Scottish Highlands and, most recently, Cornwall) and older industrial regions in need of restructuring (northern England and central Scotland, northern France, the Ruhr and Saxony areas of Germany). Increasingly, the main point for such regions has become their status in the four-yearly award of funds. Objective 1 regions, defined as those with unemployment significantly above the European Union average, received the most generous help; in the UK in the early 2000s they included areas as diverse as Cornwall, Merseyside, South Yorkshire and Wales. Many other declining regions were classed as Objective 2: they receive a lower level of support designed to aid their transition. Then, for the extended EU budget period 2007–13, the terminology changed: Objective 1 regions became Convergence Objective Regions, Objective 2 regions became Competition and Employment Objective Regions. More importantly, because in 2004 and 2007 the European Union had been hugely enlarged from 15 to no less than 27 member states, the share-out of funds spectacularly shifted from the old members to the new arrivals from Eastern Europe. Because convergence regions were defined on a strict arithmetic criterion – their gross domestic product per head must be 75 per cent or less of the all-EU average – UK regions like Merseyside and South Yorkshire lost that status, leaving just Cornwall and some thinly populated parts of rural Wales and Scotland (Figure 7.3a). In all cases the national government must provide counterpart funding – a requirement that has led to some problems with the UK Treasury. All this in turn reflects the basic fact that by the 1990s manufacturing was no longer a very significant element in UK employment, though it was rather more significant in terms of output or value added; the key was to achieve a smooth transition to the post-industrial world.

A faulty regional renaissance

With growing discontentment at the lack of UK political attention to regional economic differentials across the UK, the 1990s witnessed attempts to reintroduce a regional dimension to UK planning. In April 1994, John Major's government created a set of ten Government Offices for the Regions in England, essentially satellite offices of central government coordinating on a regional basis the activities of the Department of Trade and Industry, Department of Employment, Department of Transport and the Department for the Environment. Local government reorganization was also implemented after 1993 in England, Scotland and Wales but the form it took varied from country to country. Wales and Scotland became unitary, whereas England was treated to review area by area, with a resultant mix of new unitary and retention of existing two-tier structures in different places. And as a consequence of the implementation of the Maastricht Treaty, Scotland, Wales and Northern Ireland were designated as EU regions for the purpose of representation on the EU Committee of the Regions.

After 1997 the Labour Party was committed to regional and devolved government across the UK (Plate 5.4). Following referenda, Wales was granted a Welsh Assembly while Scotland received a Scottish Parliament. Northern Ireland was granted a Northern Ireland Assembly following the peace negotiations. London received a London-wide mayor and London Assembly. Further structural reforms were proposed for the English regions. These comprised three elements: the establishment of Regional Development Agencies (RDAs) in the ten regions of England; the establishment of a statutory form of regional planning and the preparation of Regional Spatial Strategies (RSSs), 15- to 20-year plans, in each of the English regions; and the creation of Regional Chambers or Assemblies, amalgamations of existing local authorities in each region in the interim prior to referenda determining the establishment of formal elected regional government in each region. Sub-Regional Strategies were also introduced with a statutory status to address cross-border issues. In the 2000s further local government reorganization has extended unitary status to additional areas of England, while the government has addressed major strategic or sub-regional economic, infrastructure and housing issues directly through the creation of the four Sustainable Community growth areas (Thames Gateway, Ashford, Stansted–Cambridge, Milton Keynes–South Midlands) that transcend but incorporate existing local government boundaries) and the Northern Way agenda that promotes 'city region' joint partnership working between the cities of the North and their metropolitan hinterlands (see Chapter 6 for further discussion).

Plate 5.4 The Scottish Parliament building, Edinburgh. Designed by Enric Miralles, the building opened in 2004, some five years after the implementation of devolution to Scotland. The building is meant to be a symbol of optimism for the country. The UK government's devolution programme also led to the establishment of the Welsh Assembly and the mayoral office for London and the London Assembly. The peace process in Northern Ireland led, simultaneously, to the creation of the Northern Ireland Assembly. These governmental arrangements reflect the desire for areas within the UK to determine their own policies and devise their own planning processes, suitable for their needs.

The political desire for regional government, planned for after 1997, did not materialize in the way envisaged, principally as a consequence of the government's decision to put the regional devolution issue to the public through a referendum in each region. The first test, the referendum for an elected Regional Assembly in the north of England in November 2004, was defeated by over 77 per cent of the electorate, against a turnout of 47 per cent. In strong Labour Party territory, this was an embarrassment for the government and effectively caused the abandonment of not only further elected government referenda for the remaining English regions, but also the scaling back of the regionalism agenda politically. The reasons for this result are varied; however, it was considered that regional power would have been concentrated in an assembly situated in Newcastle-upon-Tyne, which only served to strengthen strong historic rivalries between cities in the North East. But it was also unclear whether a sufficient case had been put forward for the necessity of a directly elected Regional Assembly, at a time when public scepticism has heightened towards politicians, public servants and possible increased taxation for the citizens of the areas affected. The regional agenda stalled in 2005, with aspects of regional institutionalism implemented (such as the RDAs and RSSs) but without the democratically accountable process to accompany them.

In 2007 a Treasury Sub-National Review recommended that greater powers should be given to local authorities and that the interim Regional Chambers and Assemblies should be phased out of existence by 2010. Furthermore, as part of this sub-national review, the RDAs' Regional Economic Strategies and the Regional Assemblies' Regional Spatial Strategies would be replaced with Integrated Regional Strategies. Strategic planning issues after 2010 are addressed across the UK through the Wales Spatial Plan, the Scottish National Spatial Planning Framework and the Northern Ireland Regional Development Strategy (see Figure 5.5); these are not planning documents *per se* but rather are corporate plans and have Cabinet and legislative backing. There is no UK-wide national strategic plan or national spatial framework. Rather, the UK framework has comprised all country plans for Scotland, Wales and Northern Ireland, a Spatial Development Strategy ('the London Plan') in London and the present nine RSSs in England, to be replaced with Integrated Strategies in due course. But the Conservative Party's 2010 Green Paper proposes to replace the RSSs or Integrated Strategies, with no formal intermediate level between a national spatial strategy and simplified local plans for counties and unitary areas.

Matters of national significance

Over the last 20 years, alongside attempts to deal with matters of regional concern across the UK, governments have also become more preoccupied with development of national importance and the need to resolve these matters through the planning process. Historically, major planning projects such as new railways, roads, energy plants and airports were determined either through specific parliamentary bills or through ministerial 'call-in', where planning applications raising issues of more than local interest were determined by a central government minister following a lengthy public inquiry. One of the most significant projects determined through the public inquiry process was for Terminal 5 at London Heathrow Airport, which became a *cause célèbre* in political and media viewpoints for the 46-month duration of the inquiry and the fact that the planning process took 14 years to determine at a cost of £63 million. The terminal was opened in March 2008, some 19 years after the project's inception, but the whole saga raised significant questions about not only the public inquiry process but also the role of planning and central government in shaping the decision-making system. This, in itself,

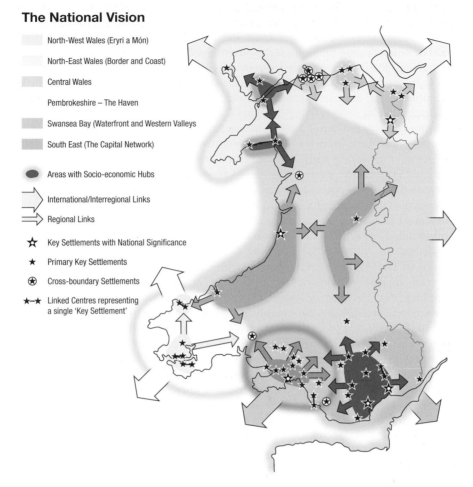

The National Vision

North-West Wales (Eryri a Môn)

North-East Wales (Border and Coast)

Central Wales

Pembrokeshire – The Haven

Swansea Bay (Waterfront and Western Valleys

South East (The Capital Network)

Areas with Socio-economic Hubs

International/Interregional Links

Regional Links

☆ Key Settlements with National Significance

★ Primary Key Settlements

✪ Cross-boundary Settlements

★–★ Linked Centres representing
a single 'Key Settlement'

Figure 5.5 Examples of a spatial development strategy in the UK post-2000: the Wales Spatial Plan 2008 national vision, building upon the original version released in 2004 by the Welsh Assembly government. This document has Cabinet-wide backing and serves as a national strategy for Wales while identifying areas within Wales that possess particular attributes that require addressing by local authorities and other agencies.

is a far cry from the way governments directly utilized the planning system between the 1930s and 1960s to impose major developments in different parts of the country, such as new towns, new motorways and nuclear power stations, without any or minimal public consultation methods. But the last 40 years has witnessed society becoming increasingly pluralist, with a concomitant shift from representative government to participatory governance, enhanced public participation and involvement mechanisms in policy development, and enhanced vocal rights within decision-making structures.

The government has provided national planning policy guidance on a range of issues of concern locally, such as flood risk, housing development, retail change and economic development, much more prominently since the late 1980s. Decisions at the local level have needed to take account of a range of these national policy statements in formulating strategies, but with little guidance as to which issues take priority or without any acknowledgement of regional differences in the severity of problems. What has been largely absent is any sense of national policy priorities or clear policy statements of projects of national significance. The reasons for this absence relate to political nervousness on the part of successive ministers. Imposing a direction in a particular substantive policy could be seen as riding roughshod over other democratically elected tiers and create major externalities for local areas, an issue made even sharper as mechanisms for land use change have embraced stakeholder participation at the grassroots level. And yet the alternative, progressing projects through a protracted public inquiry, has also proved to be unpalatable, as the long-running Terminal 5 saga demonstrated. The problems with these emerging forms of managing land use change and their interrelationships rest on a number of core issues, the most prominent of which concerns the rights and responsibilities of national governments to shape and resolve nationally significant issues (Plate 5.5). Within a changed government structure that has increasingly emphasized devolved, regional and (as we shall see in Chapter 6) local governance, the UK government has found it difficult to be seen to be leading processes to assist in action on national land use priorities. The problem of how to determine projects of national and regional significance has finally been changed recently. Under the provisions of the Planning Act 2008, the government established the Infrastructure Planning Commission (IPC) to determine development projects of major national and regional significance. Not only was this intended to be a fast track method of delivering major projects, it was significant for handing over the assessment and decision on these matters to an independent body separate to direct democratic accountability. So controversial have been the plans that the Conservative Party was committed to abolishing the IPC after it tooks office after the 2010 general election and replacing this with more project-specific parliamentary bills.

One of the key infrastructure issues that has received the most significant attention politically over the last two years has concerned the provision of high-speed rail lines across the UK. High Speed 1 was the name applied to the Channel Tunnel Rail Link between Folkestone and London St Pancras, completed in 2007. In January 2009 the government established a company to look at the principle of developing High Speed 2, a high-speed railway link from London to Birmingham with spurs to Manchester and to Leeds via the East Midlands. High-speed rail has been supported in principle by the three main UK political parties but there remains debate about which cities should be served, whether there should be a detour via London Heathrow airport and on the environmental performance and impact of high-speed rail. If approved, construction would begin in 2017 with the first trains running by 2025 and journey times to Midland and north-west cities could be reduced by 20–30 minutes.

A verdict on regional economic policies, 1960–2010

To reach an overall verdict on the impact of policies is not easy. The most weighty of the academic evaluations are far from being in agreement. Nevertheless, some fairly definite conclusions can be drawn.

The first is that over the post-Second World War period in Britain, the basic economic geography has profoundly changed. The pattern from the late 1970s onwards is best described, in Keeble's words, as a centre–periphery model. The big older conurbations were losing jobs, while peripheral, largely rural regions were gaining them: the main beneficiaries were regions like East Anglia, the South West, rural Wales, some of rural northern England, and the Highlands and islands of Scotland. Many factors played a part, but among them were equalization of location potentials because of greatly improved road transportation, a labour force perceived to be of higher quality (and lower militancy) and, simply, a better perceived residential environment. In contrast, the image of the older

Plate 5.5 St Pancras International railway station. A popular planning project that opened in 2007 at the London end of 'High Speed 1', the Channel Tunnel Rail Link between England, France and Belgium. Projects of national importance such as this one have long been subject to difficult political and planning processes. As the country will continue to require infrastructure renewal and investment for such services as new high-speed rail lines, nuclear power stations and reservoir, increasing focus will be placed on the appropriate democratic level of decision-making to determine the projects and which regions of the country will benefit directly from the new developments.

conurbations is now a profoundly negative one. As one perceptive journalist, Ann Lapping, put it: 'Any manufacturer consciously locating in London would have to be nuts.'

The second point is paradoxical: it is that regional policies have had an effect, but that sometimes it may have been unnecessary while at other times it was barely enough. Basically, regional policy did not pay sufficient attention to the profound emerging differences between the conurbations of the rural peripheries, whether in the assisted areas or elsewhere. The result was almost certainly that aid went to areas that might have flourished without it, while in the conurbations it could not offset the massive decline in employment.

The third point is that overall, regional policies have undoubtedly created jobs which would not have been located in the assisted areas at all in their absence. Though the experts differ widely in their estimates of this job creation, it is substantial and may have averaged some 40,000 jobs a year in the late 1960s and perhaps 10,000 jobs a year in the more depressed 1970s, on Moore's and Rhodes's estimates. Furthermore, since these jobs resulted in lower unemployment benefits and higher tax yields, the net cost to the Exchequer may have been negligible or zero. However, there does remain doubt as to whether the precise bundle of policy instruments has been optimal. In particular, the emphasis on subsidies to capital (through investment grants) may have been perverse when the main emphasis was on providing new jobs and reducing unemployment. Nevertheless, it must be recognized that over the period 1960–80 most of the recognized indices of regional performance, including unemployment, earnings and income, were tending to converge as between regions.

However, in this entire process the problem regions were doing little better than standing still. Though differences narrowed, they persisted, and though new jobs were created, they barely made up for those being lost. Furthermore, this effect is concealed by the conventional presentation in terms of traditional standard regions. From the mid-1970s onwards, the problem of the inner conurbation cities became increasingly serious and thus even when regional aid was available (as in Glasgow and Liverpool, but not in Birmingham or London) it was quite insufficient to stem the job loss. Hence a new emphasis – begun under Labour in the late 1970s, continued by Thatcher in the 1980s – on aid targeted to the inner cities, in the form of enterprise zones, urban development corporations, the urban programme, city grant and the like. After 1997 the Labour government embarked on a further reversal – reintroducing a regional policy dimension – but here the focus was on creating the structure and institutions of regional governance, rather than an explicit central policy of regional assistance. In fact, part of the ideology behind Labour's new regional policy stemmed from New Right thinking of the 1980s and the belief in competitiveness. Regions, it was believed, could compete economically with other world regions and indeed themselves for future investment. They would do so with 'institutional capacity building' to empower local actors to develop networks and associations to foster sustainable economic development.

The form of regional planning after 1997 has to be viewed, therefore, within a broader political and governance context. Among the drivers of change causing this regional renaissance are a political settlement through asymmetrical devolution to Northern Ireland, Wales and Scotland, and the subsequent extension of competence in these devolved territories. England, by contrast, saw a 'quiet' form of regionalism within a fluid institutional architecture that has involved new institutional structures, such as the Regional Development Agencies, Government Offices in the Regions/Regional Co-ordination Unit, the Regional Assemblies and, more recently, regional partnerships. It was always a fragile consensus.

Attention has also focused on regional impacts of national decisions within the UK, especially on major infrastructure developments and who should decide these. With a push towards enhanced public participation in planning at all levels and a reluctance of central government to be seen to be determining large projects, it became necessary to reintroduce a form of centrally led planning, even if this makes politically controversial development projects potential election issues.

Further reading

The standard works on regional policy in the 1960s and 1970s are Gerald Manners, David Keeble, Brian Rodgers and Kenneth Warren, *Regional Development in Britain* (Wiley, second edition, 1980), and David Keeble, *Industrial Location and Planning in the United Kingdom* (Methuen, 1976). Gavin McCrone, *Regional Policy in Britain* (Allen & Unwin, 1969), Harry W. Richardson, *Elements of Regional Economics* (Penguin, 1969) and A.J. Brown, *The Framework of Regional Economics in the United Kingdom* (Cambridge University Press, 1972), are also useful. David H. McKay and Andrew M. Cox, *The Politics of Urban Change*, Chapter 6 (Croom Helm, 1979) provides a useful summary of policy changes. Important evaluations are Barry Rhodes and John Moore,

'Evaluating the Effects of British Regional Economic Policy', *Economic Journal*, 33 (1973), pp. 87–110, and ibid., 'Regional Economic Policy and the Movement of Firms to Development Areas', *Economia*, 43 (1976), pp. 17–31. A valuable summary is Christopher M. Law, *British Regional Development Since World War I* (Methuen, 1981). For the period since 1980, see especially Paul Lawless and Frank Brown, *Urban Growth and Change in Britain: An Introduction* (Harper & Row, 1986); Paul Lawless and Colin Raban (eds.) *The Contemporary British City* (Harper & Row, 1986); John R. Short and Andrew Kirby (eds.), *The Human Geography of Contemporary Britain* (Macmillan, 1984); Ray Hudson and Allan Williams, *The United Kingdom* (Harper & Row, 1986); Paul N. Balchin, *Regional Policy in Britain: The North–South Divide* (Paul Chapman, 1989); John Glasson and Tim Marshall, *Regional Planning* (Routledge, 2007); and Stefanie Dühr, Claire Colomb and Vincent Nadin, *European Spatial Planning and Territorial Cooperation* (Routledge, 2010).

6 Planning for cities and city regions from 1945 to 2010

At the end of Chapter 4 we summarized some of the chief features of the elaborate planning system set up in Britain just after the Second World War. We saw that, essentially, the system was designed for an economy where the bulk of urban development and redevelopment would be carried out by public agencies – a far cry indeed from the actual world of the twenty-first century. We saw too that an essential function of the system was to control and regulate the pace and direction of change – social, economic and physical. It was assumed that control of change was both feasible and desirable: feasible, because the pace of population growth and of economic development was expected to be slow, and also because new and effective powers would be taken to control the regional balance of new industrial employment; desirable, because decision-makers generally shared the Barlow hypothesis that uncontrolled change before the war had produced undesirable results. Furthermore, we noticed that the administrative responsibility for operating the new system was lodged not in central government but in the existing units of local government, with only a degree of central monitoring. The system thus created was from the beginning more powerful on its negative side than on the side of positive initiative.

These features were of course interrelated. Because the pace of change was expected to be slow, it seemed possible to control it. Because the positive role in development would be taken by public agencies, the remaining negative powers could safely be vested in the local authorities. The danger was that if any one of the basic assumptions proved wrong, the logical interrelationships would also go wrong. And in fact the post-war reality proved very different from the assumptions of those who created the planning system between 1945 and 1952.

The reality of change in Britain

The story of Britain in the half-century after the Second World War has been one of rapid change, unparalleled in some respects during any other era, save that of the Industrial Revolution. Partly because of the speed of change, partly because of shifts in political philosophy, a much larger part of the resulting physical development has been undertaken by private enterprise than was expected in 1945 or 1947.

First, and most basically of all, the early part of this period was one of unprecedented population growth – unprecedented, at any rate, by the standards of interwar Britain. Immediately after the war there was a sudden 'baby boom': a rise in the birth rate resulting from delayed marriages and delayed decisions to have children. Demographic experts predicted both this boom, in 1945–7, and its subsequent waning, from 1948 to about 1954. By the latter date the crude birth rate (number of babies born per thousand people) was sinking towards the level of the 1930s. The experts, who were advising the planning officers that they should plan for an almost static national population total in the near future, seemed to be vindicated. But from 1955 to 1964 they were plunged into disarray by an unexpected and continuous rise in the national birth rate; by the mid-1960s the crude birth rate – 18.7 per thousand in 1964 – was threatening to approach that at the outset of the First World War. As a result the official national projections of future population had continually to be revised upwards. Whereas in 1960 the population at the end of the century was expected to be 64 million, by 1965 the projection had been raised to no less than 75 million. From this point the birth rate fell again, to 15.0 per thousand in 1972, and the projection was scaled down to 66 million. The birth rate fell even more precipitously, to a low of 11.6 in 1977, before climbing marginally to 13.0 in 1979, and the projection gave only a minimal growth of population down to the end of the century, from 54.4 million to 57.3 million. During the 1980s and 1990s the birth rate remained at between 13 and 14 per thousand, then falling to just over 12 in 1998, and the projections accordingly remained modest. Total fertility rate – a measure of population replacement, now preferred by demographers – fell precipitously from 1971 to 1977, then rose modestly in the 1980s before falling again through the 1990s; in 2003 it reached the lowest point ever of 11.2. Since then the rate has increased modestly to reach a figure of 13 in 2008. In 2009 the Office of National Statistics revised the figures and has now projected further increases in the population over the next 25 years to about 65 million in 2018 and 71 million in 2033. Figure 6.1a shows these fluctuations.

These changes have always been of critical importance to planners everywhere in the country: as will be seen in detail in later chapters, population forecasts influence almost every other forecast the planner has to make. Housebuilding programmes, projections of car ownership and demands for road space, forecasts of recreational demands and their impact on the countryside – all were automatically revised upwards. After having peaked in the 1960s at over half a million a year, by the 1990s it was of the order of only 200,000 a year, though strong internal shifts in the geography of population were leading to continued demands for new suburban development in wide rings around the conurbations. The projection was for a modest increase of around 4.4 million people in the UK by 2021 but already that has been revised upwards. (Figure 6.1b).

This growing population, furthermore, proved to be splitting itself up into an ever-increasing number of smaller and smaller households – the product of social changes such as earlier marriages, the tendency of many young people to leave home in search of educational or job opportunities, and the increasing trend for retired people to live by themselves in seaside colonies. As a result, while the average size of a home in Britain remained roughly constant, average household size almost halved in the twentieth century, from 4.6 in 1901 to 3.2 in 1951 and only 2.4 at the end of the century. (The exception consisted of certain sociocultural groups such as the Muslims in London and some northern cities, which had higher birth rates and considerably larger households.) Thus people were enjoying more space within their homes, but an ever-increasing number of homes was needed to accommodate any given number of people. Reinforcing the rise in population, this trend meant that the total housing programme, and the consequent demands on space, were much greater than had been comfortably assumed in the late 1940s.

At the same time, as we saw in Chapter 5, the population proved to be much more mobile than the planners had been assuming when they made their original local development plans after 1947. The Barlow Commission seems to have assumed, and the professional planners followed them without serious question, that it would be possible after the war largely to control inter-regional migration through effective controls on new industrial location. But as we saw, this belief was totally unjustified: there were no

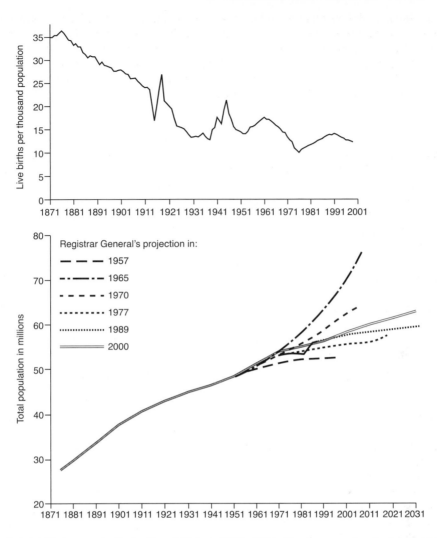

Figure 6.1 (a) Birth rates in England and Wales, 1871–2001. The long secular decline in birth rates was halted at the end of the Second World War and again by an upswing from 1955 to 1964; but afterwards births again declined. (b) Population growth in Great Britain, 1871–2031. As a result of fluctuations in birth rates, projections of future population have changed markedly in recent years.

controls on the dynamic service sector of employment, and even the controls on factory industry could be largely circumvented by repeated small-scale extensions or by buying up existing factory space. The result was a continued and strong net drift, on the lines of the movement that had so alarmed the Barlow Commission, from Scotland and Wales and northern England towards the Midlands and South. After 1960, as we shall see, this continuing trend came in for much attention and criticism. But it is worth noting that even when it was stemmed – as in the South East after about 1966 – rapid population growth continued to take place because of the strong underlying trend in natural growth within the region itself.

Coming together, these trends could only mean continuing, and even increasing, pressure for new urban development in and around the big urban areas, especially in the Midlands and South (Plate 6.1). It was in these areas, above all around London and Birmingham, that planning authorities were most taken by surprise by growth in the late 1950s and early 1960s. At the same time the natural increase of population in the development areas created a potential demand for even more employment than had been expected, increasing the scale of the problem of economic development there; though a generally expanding economy provided a steady total of mobile industry to move into these areas. In general, the pace of change created great problems for a generation of planners schooled to believe that change in itself was not particularly desirable.

Lastly, rising prosperity after 1955 resulted in rapid buying of durable consumer goods which created demands for more useable space in and around the home, interrupted only fitfully by economic setbacks like the energy crisis of 1973–74 and the deep recessions of 1980–81 and 2008–9. Above all, it was the rise in mass car ownership that perhaps took planners most by surprise. When local authorities and the first new towns drew up their development plans, austerity was still the rule of the day, and car ownership levels were barely above those of the 1930s; only about one in ten households owned a car, and new town planners felt safe in fixing one garage to four houses as a generous norm.

But by 2008, 76 per cent of households owned at least one car (against 31 per cent in 1961) and 32 per cent owned two or more; one garage per family (with additional space in reserve) had become the standard almost everywhere. Not only, however, did the new cars need houseroom in the residential areas; they put increasing strain on the country's road system, the most congested in the world, and this resulted in a constantly increasing road-building programme from the mid-1950s on. Rising car ownership was in part a response to the increasing decentralization of populations from the cores to the suburban fringes of the major urban areas, a development which was already observable in the 1930s but which gained momentum in the 1960s. But since employment and urban services (above all retailing) did not decentralize so rapidly, mass motorization created tremendous pressures for urban reconstruction in the form of urban motorways and multi-storey car parks, threatening the existing urban fabric as never before in history.

In any event, changes of this magnitude would probably have compelled a massive readjustment in the objectives, methods and machinery of planning. In particular, it is hard to see how public programmes could have adapted themselves quickly to the challenges of rising population, continuing mobility and greater affluence; private enterprise, almost inevitably, would have had to undertake a greater role than was foreseen in 1947. As a coincidence, the onset of the period of change came with the arrival of a Conservative government in 1951, heavily committed to reliance on the private sector. Very rapidly thereafter, the balance of the housing programme shifted from an emphasis on the public sector to approximate equality between the public and the private programmes. But it is significant that after its return in 1964, the Labour government did not significantly change this relationship. The mixed economy in urban development is one of the facts of life in postwar Britain.

This, in turn, had serious implications for the administrative machinery of planning. Far heavier responsibilities came to rest upon the local planning authorities, who were required to deal with a much larger amount of complex change than had ever been anticipated. And because the fundamental trends of the interwar period continued after all – population moving into the great urban regions, and simultaneously out of their congested inner areas and into their suburban fringes – the emphasis on local private initiative exposed the failure, in and just after the 1947 Planning Act, to grapple with the problem of fundamental local government reform. By loading responsibility for plan-making and development control with the separate county borough and county authorities, the Act divided cities from their hinterlands, and made effective planning of entire urban regions a virtual impossibility.

Plate 6.1 Suburban development at Heswall, Cheshire. Though after the Second World War it was at first thought that most urban development would be in comprehensively planned new towns, the unexpected population growth of the late 1950s and early 1960s – plus a changed political climate – led to significant private building investment on more conventional lines.

Planning in the 1950s: cities versus counties

It took some time for this lesson to emerge. Contemplating a more leisurely pace of change, the architects of the 1947 Act had believed that effective coordination of the different local plans could be achieved through Whitehall vetting, first in the light of the major regional advisory plans (such as Abercrombie's for London), and then with the aid of monitoring and updating by regional offices of the Ministry of Town and Country Planning (as it was then known). Ironically, soon after coming to power in 1951 the Conservatives (apart from changing the name to the Ministry of Housing and Local Government) abolished these regional offices as an economy measure. Henceforth there was no machinery for effective regional coordination, save such as could be provided from London; the various local authorities were left to stand up for themselves. At the same time, and in much the same spirit, the new government made it clear that though the existing new towns would be completed, the emphasis in future would be on voluntary agreements between local authorities to expand existing towns, within the framework of the 1952 Town Development Act.

Furthermore, the government made it clear that it favoured a very negative attitude towards urban growth. Encouraged by the falling birth rates of the early 1950s, in a famous Circular of 1955 it actively encouraged the county authorities to make plans for green belts around the major conurbations and freestanding cities; and subsequently the minister indicated that even if it was neither green nor particularly attractive scenically, the major function of the green belt was simply to stop further urban development (Plate 6.2). And the green belts were only one element in a series of planned constraints on Greenfield development, including also national parks and areas of outstanding natural beauty (AONBs) (Figure 6.2). But since there was no effective machinery for coordinated regional planning, and since the 1952 Act was proving ineffectual because of weak financial provisions, these strong negative powers were not accompanied by any positive machinery for accommodating the resulting decentralization of people from the cities and the conurbations.

The result, in the late 1950s, was a series of epic planning battles between the great conurbations and their neighbouring counties, culminating in a number of major planning inquiries – notably those on the proposals of Manchester to build new towns at Lymm and at Mobberley near Knutsford in Cheshire, and on the proposals of Birmingham to develop at Wythall in the Worcestershire green belt. In a series of contests on both these

Plate 6.2 A view of the London green belt at Cockfosters, North London. The effectiveness of the green belt is well illustrated by the sudden stop to London's urban area. Most of the land in this picture would almost certainly have been developed but for the postwar planning controls. Trent Park, on the right of the picture, is a country park designated under the 1968 Countryside Act.

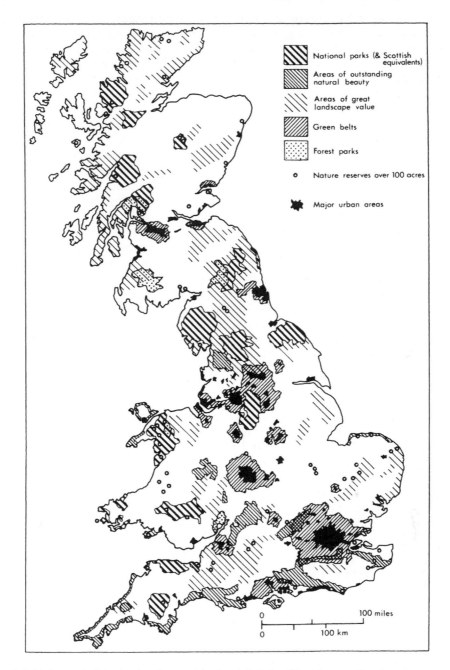

Figure 6.2 Major restraints to development in Great Britain. Altogether, national parks, green belts and other specially designated areas exclude more than 40 per cent of Britain's land area from the prospect of large-scale urban development of any kind. Little of this land had any protection before the Second World War.

cases, the cities lost and the counties preserved their rural acres. In every case, the overriding need to conserve agricultural land was quoted, echoing the words of the 1942 Scott report. But in the 1958 Lymm inquiry, the agricultural economist Gerald Wibberley produced powerful evidence to indicate that the agricultural value of the land involved was fairly low compared with the extra costs which might be involved in high-density redevelopment within the cities. At about the same time another economist, Peter Stone, was beginning to demonstrate just how great these additional costs of high-density redevelopment could be. Yet government subsidies continued to encourage such high-density schemes.

By the end of the 1950s the position was becoming desperate for the cities. They had been encouraged in 1955 to start again on their big slum-clearance programmes, which had been interrupted at the start of the war in 1939; but ironically, this came at just the point when the birth-rate rise began (Plate 6.3). Adding together the demands from rising population, household fission, slum clearance and overcrowded families, and reporting on their virtual failure to get substantial agreements under the 1952 Act, city after city by 1960 was simply running out of land for its essential housing programme. The plight of the cities was underlined in this year in an influential book by Barry Cullingworth, *Housing Needs and Planning Policy*. In the same year Geoffrey Powell, a government official, pointed out publicly the rapid rate of population growth in the ring of Home Counties around London – an increase that no one, official or otherwise, would have thought likely ten years earlier. A year later, the census showed that population growth in this ring during the 1950s had amounted to 800,000 people – one third of the net growth of population in Britain. And contrary to the expectations of the architects of the 1947 system, the vast majority had been housed not in planned new or expanded towns, but in privately built suburban estates on the familiar interwar model. The only change was that the green belt had been held; the developers had therefore leapfrogged it, pushing the zone of rapid population growth into a wide band up to 50 miles from the centre of London (Figure 6.3). And, over the succeeding decades, this urban frontier would move progressively farther out (Figure 6.4).

The changed situation could not be ignored. In 1961 the government, reversing its previously declared policy, announced the designation of the first new town in England for 12 years: Skelmersdale near Wigan in Lancashire, designed to relieve the pressing needs for housing of the Merseyside conurbation, which had the most concentrated slum clearance problem in England. In the same year, in an even more fundamental reversal, the government began once again to embark on a series of major regional planning studies, intended to provide guidelines for the plans of the individual local planning offices. A new era for planning had begun.

The major regional studies of the 1960s

Shortly after the government had embarked on the first of these studies, focused on the South East, regional planning received a sharp impetus from a different source: the recession of 1962–3, which – as we already saw in Chapter 5 – led to the production of stop-gap regional development plans, on what was virtually an emergency basis, for the distressed regions of the North East and central Scotland. As at the time of the Barlow report in 1940, two main strands in British regional policy again fused: one, the objective of more rapid economic development in the development areas; the other, the attempt to control and channel the rapid growth around the more prosperous major conurbations, such as London and the West Midlands. The first of these strands has already been discussed in Chapter 5, where we saw the evolution of a much stronger policy of regional controls and incentives during the 1960s. It is the second that is relevant here.

The South East Study, published in March 1964, caused considerable surprise and even political controversy by its major conclusions: that even with some measures to restrain the further growth of the region around London, the pressures for continued expansion were such that provision must be made to house a further 3.5 million people in the whole region during the 20-year period 1961–81. The main justification for this, which few critics seemed to realize at the time, was that even then the main reason for the population growth of the region was not the much publicized 'drift south' of able-bodied workers from the North and from Scotland; the reasons were the natural growth

Plate 6.3 Reconstruction in London's East End. Formerly the scene of some of London's worst slums, the East End has been largely reconstructed since the Second World War. The old terrace houses have been replaced by mixed development including tall blocks of flats – now criticized, here as elsewhere, on social and aesthetic grounds. Rebuilding could not house all the former population, so some left London in planned dispersion under the 1944 Abercrombie Plan.

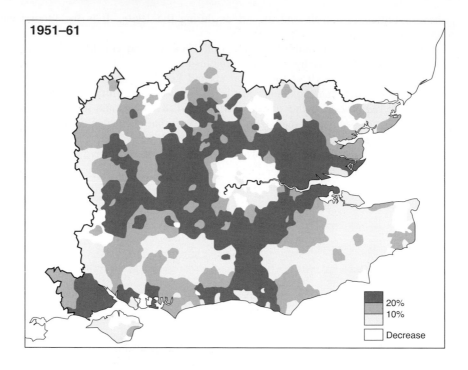

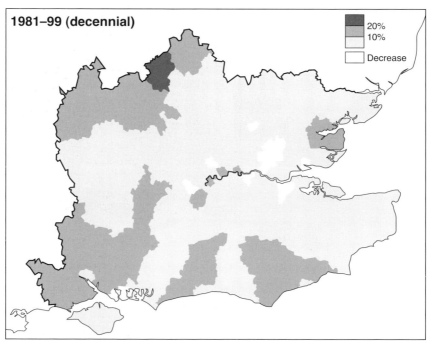

Figure 6.3 Population growth in the London region, 1951–61 and 1981–99. Since the Second World War Greater London has become a zone of widespread population loss, surrounded by a belt of rapid gain which has moved steadily outward. By the 1980s and 1990s the fastest gains were being recorded 40 and more miles (65-plus kilometres) from London. Similar patterns of urban decentralization were recorded around other conurbations, though on a smaller scale.

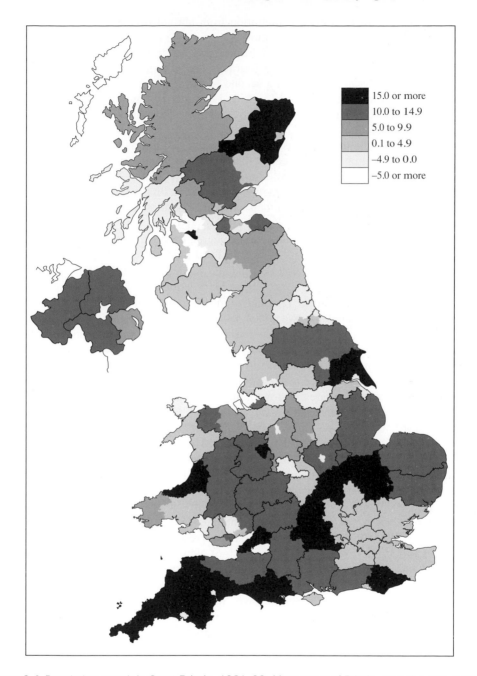

Figure 6.4 Population growth in Great Britain, 1981–99. Most areas of Britain, save the conurbations and older industrial areas, have gained population. But particularly clear is the loss from the major conurbations to their suburban fringes, especially around London. Rapid population growth characterizes wide rural areas of the country. Particularly noticeable is the 'golden belt' at the borders of the South East region with the South West, the Midlands and East Anglia.

of the region's own population, migration from abroad (much of which had been cut off by the Commonwealth Immigration Act of 1962) and migration of retired people to the South Coast resorts. To channel the pressures for growth, and to avoid the problems of congestion and long commuting journeys focused on central London, the report recommended a strategy based on a second round of new towns for London at greater distances than the first round, well outside London's commuter range: Milton Keynes in northern Buckinghamshire, 49 miles (80 kilometres) from London; Northampton, 70 miles (110 kilometres) from London; Peterborough, 81 miles (130 kilometres) from London; and Southampton–Portsmouth, 77 miles (120 kilometres) from London, were among the more important projects which finally went ahead in one form or another (see Figure 6.5 and Plates 6.4a, b). Significantly, only Milton Keynes among these was a greenfield new town on the old model, albeit with a bigger population target than any previously designated town; the others all represented a new departure in being new towns attached to major existing towns or cities. To cope with the development nearer London which might still depend to some degree on commuting, the study was much less clear; this, it implied, was a matter for local planners. The Labour Party, returned to office in October 1964, at first demanded a second look at the study's conclusions; later, it accepted them in large measure.

Meanwhile, in 1965 further regional studies, produced in the same way by official *ad hoc* teams, appeared for two other major urban regions: the West Midlands and the North West. Recognizing the changed situation brought about by population growth and by the continuing slum clearance programme, both studies calculated that the problems of accommodating planned overspill from the conurbations were greater than had earlier been appreciated – even though the North West was an area of net out-migration. To accommodate the growth, the reports called for an accelerated programme of new town building in each region, neither of which had received any new towns before 1961. By 1965 two new towns had already been designated in the Midlands: Dawley (1963) and Redditch (1964); later (1968), as the result of the study's conclusions, Dawley was further expanded and renamed Telford. In the North West, Skelmersdale (1961) and Runcorn (1964) had been established to receive Merseyside overspill; the report suggested that similar developments were needed for Manchester, and this need was eventually met by designations at Warrington (1968) and Central Lancashire (Preston–Leyland, 1970). In central Scotland, further overspill pressures from the central Clydeside conurbation (Greater Glasgow) resulted in the designation of Livingston (1962) and Irvine (1966). And lastly, south of Newcastle the new town of Washington (1964) was designated to receive overspill from the Tyneside and Wearside areas. Thus, in addition to the new towns for London, the major conurbations outside the capital had no less than nine new towns designated between 1961 and 1970 to aid with their overspill problems (see Figure 6.5). Significantly, as with the London towns, several of these – notably Warrington and Central Lancashire – worked on the formula of attaching a new town to an old-established existing town possessing a full range of urban services.

These reports of 1963–5 mark an important stage in the evolution of British postwar urban planning. For they recognized officially that the fact of continued population growth demanded positive regional strategies, covering areas that embraced the conurbations and a wide area around them; the reports themselves make it clear that this wider area extends much farther than the conventional 'sphere of influence' defined by geographers in terms of commuting or shopping patterns, and may even in some cases approximate to the area of the wider region used for purposes of economic development planning. These strategies, involving new and expanded towns, were required

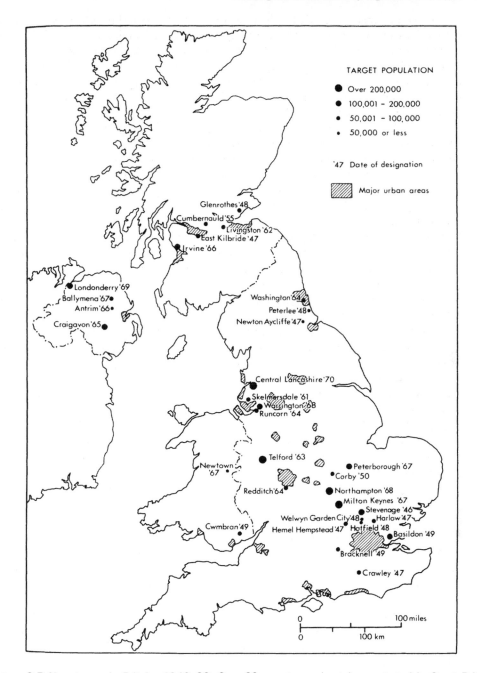

Figure 6.5 New towns in Britain, 1946–80. Over 30 new towns have been started in Great Britain and Northern Ireland under the 1946 New Towns Act and its Northern Irish equivalent. They fall naturally into two groups: Mark 1 new towns of the 1946–50 period, concentrated around London and in the development areas, and a second wave started in the 1960s to serve the needs of the major conurbations.

Plate 6.4 (a) Milton Keynes new city, Buckinghamshire. Designed in the late 1960s, this was planned as a series of low-density housing areas in the interstices of a rectangular highway grid. The aim was to promote easy mobility both by private car and by public transport. It has been more successful in the former than the latter.

even for those conurbations in the development areas; there, despite continued out-migration, slum clearance and natural growth necessitated a positive overspill policy. Essentially, what these reports represented was an application to the major urban regions around London of the standard Howard–Barlow–Abercrombie formula: planned decentralization of the conurbations, coupled with green belt restrictions and new communities placed in general outside the commuting range of the conurbations. Save for an emphasis on the housing problem, there was little attention to social policy planning; these were still physical plans in a traditional British mould.

Lastly, it is significant that these regional reports were the work of *ad hoc* teams of central government officials. This was because there was no other machinery to produce them. In 1965 the regional organization of the central government planning ministry had not been restored. (This took until the early 1970s, when a regional organization was announced for the new Department of the Environment.) Nor was there a general tradition of cooperation among the local planning authorities in each region, though some areas, notably London and the surrounding authorities, had taken a lead in the Standing Conference on London and South East Regional Planning.

Plate 6.4 (b) Cumbernauld new town, North Lanarkshire, Scotland. Designed in the mid-1950s, this is a celebrated example of a compact new town built at higher densities than the Mark 1 examples, with an extensive network of high-capacity roads. The town was intended to house overspill arising from Glasgow's slum clearance programme, and to serve development in central Scotland by attracting new industry.

The new regional structure and local government reform, 1965–72

During 1965–6 this gap in the formal machinery was partially filled. We have already seen, in Chapter 5, that the economic planning regional councils and boards were originally intended specifically to work on a rather different sort of planning: economic development planning in its relation to the national economic plan. But almost as soon as they were set up, the councils found themselves immersed in spatial planning at the scale of the city region, using that last term in its broadest sense. The logic should have been evident. Even in the development areas it was impossible to produce a development strategy for an area like the North East of England without a physical component including elements like main roads, major new industrial areas and associated housing schemes, ports and airports. And in the prosperous and rapidly growing areas, such as the South East or the West Midlands, economic planning would largely consist of a

rapidly. Employment decentralized more slowly, and during the 1950s many cities and conurbations were actually gaining new concentrations of tertiary (service) industry in their central areas; but during the 1960s there was clear evidence that this process too was being reversed. Nevertheless, in general the result of urban decentralization was increasing long-distance interdependence of the different parts of the big urban regions. Journeys to work or to shop increased in average length. The new towns programme and the expanded towns programme between them did not contribute more than about 3 per cent of the total housing programme, so the fond hope of Ebenezer Howard and the 1945 Reith Committee – that urban populations would decentralize towards self-contained communities – was never fulfilled. But as car ownership produced greater personal mobility, this seemed less necessary. Nevertheless the resulting pattern of movement created unprecedented new problems – above all for transportation planning.

By the early 1960s it was already being realized that the realities of long-term transportation planning, to accommodate projected future traffic movement, could only be handled within the framework of a wide urban region, so defined as to include both the origins and the destinations of the great majority of all the traffic movement. Following the pattern of the pioneer American transportation studies, wide-area studies began to be commissioned in Britain, first in 1961 for London, then for other major urban areas such as the West Midlands, Merseyside, south-east Lancashire–north-east Cheshire (Greater Manchester) and Greater Glasgow. But because there was no framework of local government at this level, like the major regional studies they had to be set up on a purely *ad hoc* basis. Even the reform of London local government in an Act of 1963, which created the Greater London Council, failed to encompass the entire area needed for meaningful transportation planning. (It excluded, for instance, the line of the planned orbital motorway for London.) As the various studies proceeded from fact-gathering to analysis and projection and then to proposals, the inadequacies of the existing fragmented local government structure became more and more evident. More and more *ad hoc* arrangements proved to be necessary – for instance, the passenger transport executives for the major conurbations and their fringe areas, set up under the 1968 Transport Act. The case for fundamental reform became more urgent.

It was given a further impetus in 1965 by the publication of an important report of an official advisory committee, the Planning Advisory Group (PAG). This concluded that the style of development planning set up under the 1947 Act, with its emphasis on detailed statements of future land-use proposals, did not suit the rapidly changing situation of the 1960s. Instead, there should be a new two-tier system of plan-making: first, structure plans containing main policy proposals in broad outline for a wide stretch of territory; second, local plans for smaller areas which would be prepared within the framework of the structure plans as occasion arose, including action-area plans for specific developments. (The structure plans still would be submitted for detailed vetting to the central planning ministry; the local plans in general would not.) The logic of this argument, which was generally accepted and embodied in a Planning Act of 1968, made the case for local government reform even more compelling. For the structure plans could by definition be prepared only for large areas encompassing the whole extended sphere of influence of a city or a conurbation. From 1966 onwards, some local authorities began to cooperate on an *ad hoc* basis to produce early experiments in such planning for areas like Leicester city and Leicestershire, Derby–Nottingham–Derbyshire–Nottinghamshire, Coventry–Solihull–Warwickshire, and South Hampshire (Figure 6.7). And by 1970–1, followed this lead and anticipating local government reform, most local authorities in the country were beginning to band together on city regional lines to work on the new structure plans.

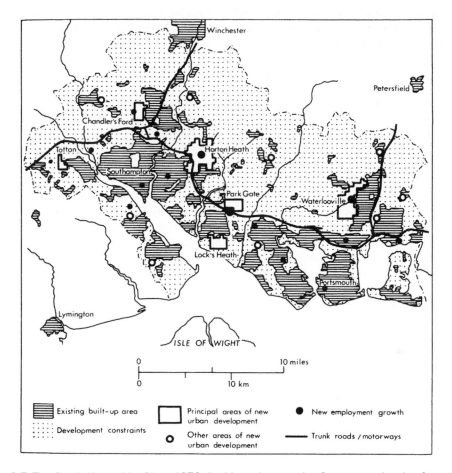

Figure 6.7 The South Hampshire Plan, 1972. In this early example of structure planning for a whole city region, major growth is grouped into a number of new communities of different sizes, close to existing urban areas and well served by public transport. Employment growth would occur both in these new areas and in the cores of the existing cities and towns.

In 1966 the government of the day – a Labour government, under Harold Wilson – recognized the logic of the new situation. It set up Royal Commissions on Local Government for England, under Sir John Redcliffe Maud (later Lord Redcliffe Maud), and for Scotland, under Mr Justice Wheatley, with a separate inquiry for Wales. Unlike a previous Commission on Local Government for England, whose terms of reference had been narrowly circumscribed, the new inquiries were specifically charged to take a fresh look at the problem. From the beginning, it was generally accepted by their members, and by the informed public, that something loosely called the city region – that is, the city or conurbation plus its sphere of influence – would be the right basis for local government reform. This indeed was the burden of the evidence submitted in 1967 by the Ministry of Housing and Local Government to the English commissioners.

The difficulty was that in practice it was more difficult than had been thought to define the city-region concept. Essentially, the new structure of local government units must possess four attributes. It must be able to perform local government services efficiently (that is to say economically, in terms of resources) and effectively (in terms of reaching the clients who need the services). It should express some communal consciousness; that is, it should take in an area which people recognize that they belong to. And it should take in the whole area whose planning problems need to be analysed and resolved together. Unfortunately, these four requirements by no means lead to a common solution. In thinly populated rural areas, efficiency may suggest big units, effectiveness small ones. The unit of communal consciousness – broadly, the area within which people travel to work or shop – may be much smaller than the planning region, which may have to take in distant sites for potential new towns.

Faced with these contradictions, the commissioners came up with two entirely different solutions when they reported, almost simultaneously, in the summer of 1969. The first was accepted by all except one of the English commissioners. It held that efficiency was most important, and that this suggested a large average size of unit; that effectiveness and community demanded the same set of units for all services; and that planning problems required to be solved by rather large units. Some compromise between these principles had to be made; the English commissioners settled on a pattern of unitary authorities capable of running all local services, covering the whole of provincial England except for the three biggest conurbations of Greater Manchester, Greater Liverpool and Greater Birmingham. Here, there would be a two-tier structure with a metropolitan authority responsible for overall physical and transport planning, and metropolitan districts for the more personal services (Figure 6.8a). This was the structure already adopted in the 1963 Act for London, which was left outside the English commissioners' terms of reference.

The opposite view was expressed by one English commissioner (Mr Derek Senior) and by a majority of the Scottish commissioners. It started from the premise that planning demanded a quite different scale from the personal services. Thus there should be a two-tier system over the whole country, based on large city regions (in the Scottish report called provinces) at the top level and on small districts at the lower level (Figure 6.8c). The English majority admitted the force of this argument, but thought that the claim of simplicity in the structure, with only one level, was overriding.

In 1970 the Labour government accepted the different prescriptions for the two countries, with minor modifications; in 1971 the Conservative government reaffirmed its broad acceptance of the Wheatley recommendations for Scotland, but replaced the English proposals by a two-tier system over the whole country. This, however, was not Mr Senior's minority prescription, but a reform based on the existing county structure at the top-tier level (with some modifications to take account of the city regional principle) and on amalgamations of existing county district authorities at the lower level (Figure 6.8b). In the conurbations the reform retained the metropolitan principle, and extended it to west and south Yorkshire, but it cut back the boundaries approximately to the physical limits of the built-up area; the green belts, and the growing suburbs beyond them, were generally left under non-metropolitan county control.

From the point of view of planning, this reorganization – which was implemented in an Act for England in 1972, and introduced the new system in 1974 – only underlined the fundamental problem of coordinating the structure-planning process over a wide area. The Redcliffe Maud proposals, by giving overall planning powers to very broadly based metropolitan authorities, might just have coped with this problem. But even they failed to take in the whole area which needed to be planned as a unit around the biggest conurbations; indeed, in the extreme case – that of London – such a unit would take in the whole of the economic planning region. The 1972 changes certainly did not aim to do this, and they merely underlined the need for some intermediate level of regional planning between the reformed structure of local government and

Whitehall. Indeed, it was clear that in many critical fast-growing areas around the conurbations – such as the Coventry area or the Reading–Aldershot–Basingstoke area west of London, designated as one of the major growth areas in the *Strategic Plan for the South East* – the newly formed local government units would immediately have to come together on an *ad hoc* basis for city-region planning, as well as being involved

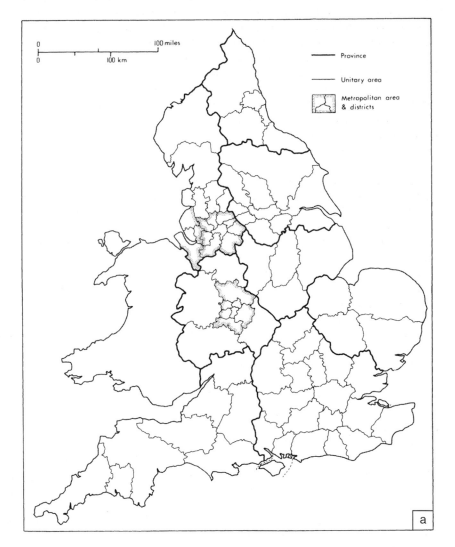

Figure 6.8 Local government reform proposals, 1969, 1972 and 1990s. (a) The English Royal Commission (Redcliffe Maud) proposals of 1969 suggested single-tier, unitary authorities for most of the country, with a two-tier solution reserved for three metropolitan areas based on the conurbations. (b) The Local Government Act, 1972, in contrast introduced a two-tier system everywhere, but with a different distribution of functions in the metropolitan counties – now increased to six – as compared with elsewhere. (c) In Scotland the Royal Commission (Wheatley) proposals of 1969 envisaged a two-tier system different from anything proposed in England, with top-tier authorities covering wide regions; this proposal, with minor modifications, was implemented but abandoned in the 1990s. (d) Reorganization in the 1990s and 2000s left the local government map even more fragmented and incoherent – a return to the 1950s and 1960s.

continued . . .

in a cooperative planning process for the wider West Midlands or South East region. The Redcliffe Maud report had recommended a structure of provincial units for England, above the unitary and metropolitan authorities, for just this purpose; but nothing came to pass from it.

The 1972–4 reorganization for England, then, was in many ways a second-best solution: it failed to recognize the realities of contemporary urban geography, and by instituting two tiers of government it created the powerful, often contraposed, planning bureaucracies. The most that could be said for it was that it recognized local interests, and that it followed traditional county boundaries in the great majority of case. But once established, it precluded further fundamental reorganization for many years. During the 1990s a Conservative government carried through a totally ad hoc reorganization of local government separately in England, Scotland and Wales, even though the case for radical reform had not been made. In Wales and Scotland, the processes were undertaken

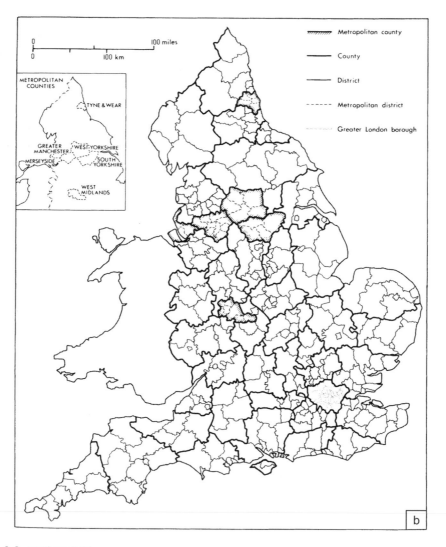

Figure 6.8 continued (b)

by the relevant government departments and the result after much intense debate within the countries was the Local Government (Wales) Act 1994 which created 22 all-purpose Welsh unitary authorities and the Local Government etc. (Scotland) Act 1997 which created 32 Scottish unitary councils. The situation in England was more complex. The government established an independent Local Government Commission (1992–2002) to oversee the case for reform in each part of England. The government stated in its advice to the commission that there was no need for consistency and that local wishes were uppermost. It rather predictably obtained a strange patchwork quilt of local government, with larger cities and towns breaking away as all-purpose unitary authorities (as with the old county boroughs before 1972–4), the abolition of two creations of the 1970s reorganization, the counties of Avon and Cleveland, and even the complete dismemberment of counties such as Somerset, Cheshire and the Royal County of Berkshire after more than a thousand years of existence (Figure 6.8d). Meanwhile, unitary status was also created in several provincial cities and large towns in 1997–8, and the government extended the reform process further and implemented unitary status in Wiltshire, County Durham, Northumberland, Shropshire and Cornwall in 2009 (in favour of the county at the expense of the districts). Respect for tradition, it seems, has its limits, such that England now has a somewhat disjointed and incremental structure of local government resting on a mix of unitary and two-tier systems.

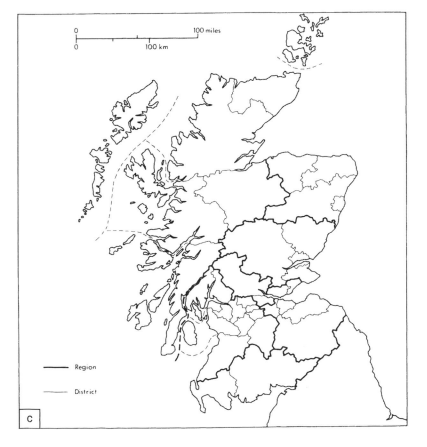

Figure 6.8 continued (c)

until environmentally acceptable limits were reached (Figure 6.11a, b). The report, which generated immense public attention and approval, took an approach which proved extremely difficult to fit into the new cost–benefit framework of the economists. When at last a major cost benefit study was undertaken for an official inquiry which depended centrally on the evaluation of environmental quality – the study of the location of London's third airport, in 1968–70 – the result was widespread public criticism and controversy. By the start of the 1970s, this was yet another important area where the techniques of planning still urgently awaited development, though at central government level the creation in 1970 of the Department of the Environment – at last integrating urban and transport planning in one organization – provided a better framework for incorporating environmental factors into decisions on transport investments.

Increasing concern with the quality of environment expressed itself in other ways too. In 1968 the Countryside Act marked an important stage in the evolution of planning for outdoor recreation. It replaced the National Parks Commission, which had been created by the 1949 Act (Chapter 4), by a Countryside Commission with wider respons-ibilities and greater financial powers. After the 1968 reform, the government promised more money to the new commission to back up these powers, though the sums still remained puny in comparison with other spending programmes. And it empowered local authorities or private agencies to create country parks near major centres of population, with subsidies from the commission, so as to act as 'honeypots' relieving pressure on the national parks (Figure 6.12). Later on, in the 1972 reform of English local government, an important change was made in the administration of national parks; though they would still be managed by the county councils, or by joint boards of county councils, henceforth there must be a separate committee and planning officer for each park. Here the government was recognizing the weighty criticism that in many of the parks the local administration had shown little enthusiasm for positive planning and hardly any willingness to spend money. The consequences were becoming serious, as increasing car ownership and motorway extensions brought floods of motor traffic into the heart of the parks at peak holiday periods. In Wales, the commission lasted until 1991 when it became the Nature Conservancy Council and later still the Countryside Council for Wales; in England, the commission became the Countryside Agency in 1999 and in 2006 was merged into Natural England.

The early 1970s: limits to growth

In 1972, with the United Nations Stockholm Conference on the Human Environment and the publication of the immensely influential Club of Rome report *The Limits to Growth*, environmental quality became a major political issue throughout the world. Britain took a number of important steps at this time. It followed up the highly successful 1956 Clean Air Act – whereby local authorities, with central government financial aid, were empowered to introduce clean air zones in Britain's towns and cities – with a radical reorganization of water-supply and sewage-disposal services, coupled with a programme that promised to clean up the country's more grossly polluted industrial rivers by the early 1980s. And, following recommendations by official committees, it reorganized compensation and road-building procedure so as to give better guarantees of environmental quality to those living alongside new highways. It provided grants for double-glazing and similar measures – a scheme already operated for householders around London's Heathrow airport; it provided more generous compensation for those wishing to move away from highway construction; and it provided for more environ-mentally sensitive, and more expensive, designs for new urban roads.

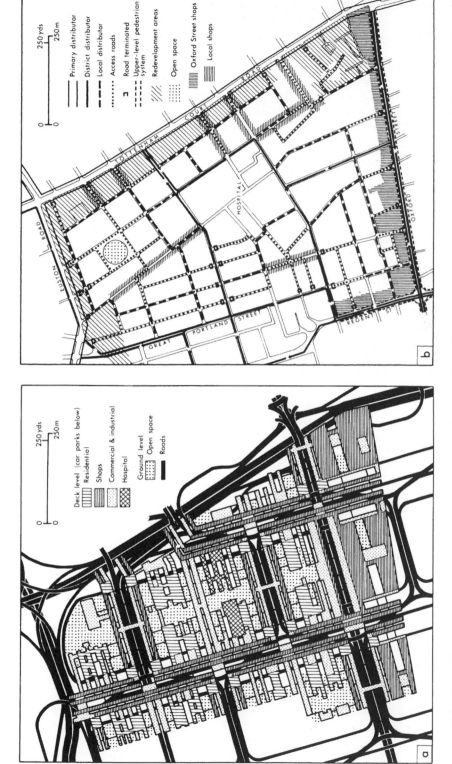

Figure 6.11 Alternative levels of urban redevelopment in part of London's West End, from the Buchanan Report: (a) maximal; (b) minimal. The report *Traffic in Towns* (1963), by a group under Sir Colin Buchanan, argued that the traffic capacity of a town should be fixed at an environmentally acceptable level. More traffic could be accommodated by complex reconstruction of the urban fabric, but this would be expensive; the alternative would be traffic restraint.

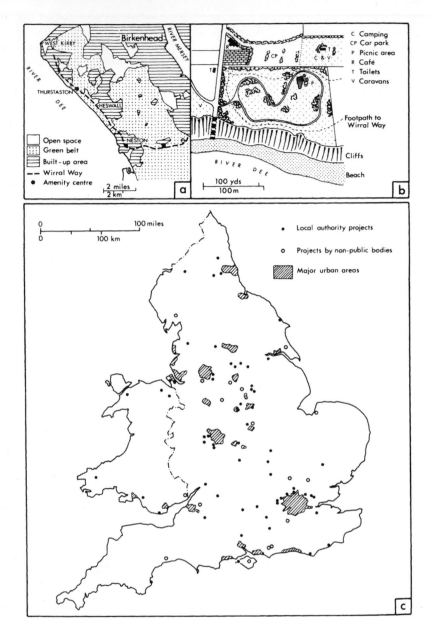

Figure 6.12 Country parks: (a) Wirral Way Country Park, Cheshire; (b) a picnic place on Wirral Way; (c) location of country parks in England and Wales, 1972. The country park concept, introduced in the Countryside Act 1968, allowed local authorities and other bodies to develop sites for intensive outdoor recreation close to the major urban areas.

All this represented an answer to a very evident demand on the part of the public – or at least the vocal section of the public. Controversies like the location of London's third airport, or London's proposed motorway system, or new roads in places like Winchester[1] and the Lake District, all demonstrated that substantial numbers of people were now very sensitive to any threat to their own environment. Their interest gave added point to the problem of incorporating environmental assessments into the evaluation of alternative plans. But at the same time it emphasized the point that in any planning decision there were likely to be winners and losers. If better-informed, better-organized groups campaigned successfully in their own environmental interests, the real risk was that the decision would go against the less informed and the less organized – who, in general, were also the poorer members of the community.

Meanwhile, two quite separate but related developments had been occurring in other fields, whose consequences for planning seemed likely to be momentous. The first was the growing concern – first evident, in the late 1960s, in American city planning but then increasingly imported into Britain – for the social objectives of planning. Essentially the argument, as developed in the United States, was that physical or spatial planning had failed many of the people that it ought to have helped, because it had not started from sufficiently clear and explicitly social objectives. In particular, critics pointed to the many examples where American urban renewal had simply displaced low-income

Plate 6.5 Elvaston Castle, Derbyshire. This country house and grounds, situated between Derby and Nottingham in open countryside, was one of the first local authority country parks designated under the 1968 Act.

residents from inner urban areas without providing alternative housing, leaving them worse off than before; and to the way in which public programmes, such as the Interstate Highway Program, had contributed to suburban dispersal of people and employment, leaving low-income inner-city residents increasingly separated from job opportunities. Great concern developed all over the United States during the 1960s at the increasing polarization of the metropolitan areas, whereby higher-income residents and their associated services and jobs migrated to far-flung suburbs, while the older central cities were left to cater for low-income residents with a constantly declining local tax base. By the early 1970s, in the inquiry on the Greater London Development Plan, fears were being expressed that a similar fate was overtaking London – and, by extension, perhaps other British cities as well.

Some of the conclusions to be drawn from such analyses were purely in terms of changed machinery: larger units of local government uniting cities and suburbs, for instance, or new sources of local revenue for cities, or revenue-sharing between central (or regional) and local governments. But more deeply, the debate seemed likely to shift the central focus of what spatial planners did. While the injection of transportation planning in the early and mid-1960s had led to an emphasis on economic efficiency as the central objective of planning, the injection of social planning in the late 1960s and early 1970s seemed certain to lead to an emphasis on equity in distribution. Planning, some sociologists were increasingly arguing, essentially distributed public goods (i.e. goods which could not be bought and sold in the market, such as clean air or quiet residential areas) to different groups of the population. It could be progressive in its social consequences, by distributing more of these goods to the lower-income groups, or the reverse. Too often, certainly in the United States but perhaps also in Britain, it had been regressive. The question that needed to be asked now was: who benefited from planning policies like urban containment, or green belts, or high-rise urban renewal? And who, conversely, suffered the disbenefits? The conclusions could be disturbing. But again, this was an item on the agenda for future development at the beginning of the 1970s.

Lastly, but relatedly, the late 1960s had seen an increasing influence on British local authorities generally of modern management techniques originally developed within private profit-making industry. The critical events here were first the publication in 1967 of the Maud Committee report (not to be confused with the Redcliffe Maud Commission report) on local authority management, which recommended a new structure for local government based on a few major committees covering wide policy areas, together with a central policy-making committee; and second, the parallel reform of the personal social services of local government during 1969–70, which created new combined social services departments embracing public health and child care. But intellectually the new movement was associated with the influence of the new techniques of Planning–Programming–Budgeting Systems (PPBS) imported from the United States and originally developed for a very different area of public enterprise: Robert McNamara's Department of Defense. Essentially, PPBS demanded that management of public enterprises should be restructured on the basis of objectives rather than of traditional departmental responsibilities. Applied to American defence problems, it asked for instance how to achieve a specific objective – for instance, how to provide for defence against surprise attack – rather than how to develop specific programmes for the army or the air force. Applied to the very different world of British local government, it would again ask how to achieve an objective – for instance, preventing the break-up of families – rather than emphasizing separate programmes of housing or child care.

In the early 1970s the influence on planning seemed likely to be profound. Many of the central objectives which local government seemed certain to develop for itself under the new PPBS framework would be social objectives: objectives in terms of people and their needs, rather than in terms of physical policies. Management by objectives, therefore, seemed likely to shift planning further away from its old emphasis on physical policies, and towards a style in which policies had to be developed, and then defended, in terms of their specific implications for the welfare of the people involved. But this in general proved a false hope. The 1972–4 reorganization, by dividing local government into two tiers, made it conceptually impossible for any local government to consider the delivery of all services against common objectives. But more fundamentally, the belief in the wisdom of the technocratic philosopher-planners had been shattered.

The first signs of this came with the work of the Community Development Plan teams, appointed by the Home Office, in a number of deprived urban areas in the period 1972–6. Their role was to stimulate the people of their areas out of apathy and into greater control over changing the conditions of their own lives. Their achievements, in many cases, were to come in violent conflict with local bureaucracies and councillors as they adopted an uncompromisingly fundamental Marxist-style explanation of the plight of their inner-city clienteles. Though the whole experiment was hastily shut down in 1976, this style of analysis had a radical influence on the approach of academic urban researchers, who increasingly analysed the woes of decaying inner-city areas in such terms.

The late 1970s: the emerging inner-city problem

Even non-Marxists, however, were coming to appreciate that there was a new dimension to planning for such areas. By the mid-1970s it was clear that they were losing people and jobs at a massive rate. Typical inner-city areas lost between 16 and 20 per cent of their populations in the 1961–71 period, and this trend continued unabated down to the late 1970s. London lost some 400,000 manufacturing jobs between 1961 and 1975 alone – and nearly 800,000 over the period 1961–84. Furthermore, as careful work by both academics and government researchers concluded, much of this job loss represented not outward movement but plant closure and, to some extent, rationalization accompanied by productivity gains. Radical researchers demonstrated this as partly representing a major restructuring of industry, accompanying business takeovers and concentrated in older inner-city firms. Skilled jobs tended to disappear in the process, but in any case there were too few jobs of any kind to go round, while much of the remaining inner-city workforce was under-educated, under-skilled and least likely to find employment in a bleak competitive world.

The result was the development of pockets of multiple deprivation in the older inner cities. Though a majority of people there were not deprived – and a majority of the deprived were to be found elsewhere – the argument remained that in concentrating resources on inner cities the government could attack the most stubborn concentrations of unemployment, poverty and want.

In 1977 the government acted on this conclusion. Accepting the results of a five-year study of three British cities (Liverpool, Birmingham and Lambeth in London) that had agreed on their broad conclusions, it announced a greatly extended Urban Programme of special aid to inner-city areas, concentrated especially on seven so-called partnership areas and 15 smaller programme areas mainly in inner conurbation cities. The resources went on a great variety of schemes aimed at job creation, winning development, sport and recreation. The Conservatives, winning power at Westminster in 1979, made no attempt to change these policies, though they did decidedly shift resources away from inner-city areas towards the suburbs, as well as trimming areal expenditure overall.

Most importantly, as already noted in Chapter 5, they introduced two new inner-city initiatives. Urban development corporations, modelled on the successful new-town formula, virtually supplanted local government in the semi-derelict London and Liverpool docklands, and later in a number of other critical inner-city areas; and enterprise zones, a somewhat watered-down version of a scheme to remove all kinds of restrictive control from the worst-hit areas, offered a combination of special tax relief and simplified planning controls in nearly a dozen decayed areas (Figure 6.13).

Their individual effects can be shown to have transformed derelict industrial areas in the heart of British cities with the creation of modern offices, apartments, commercial and

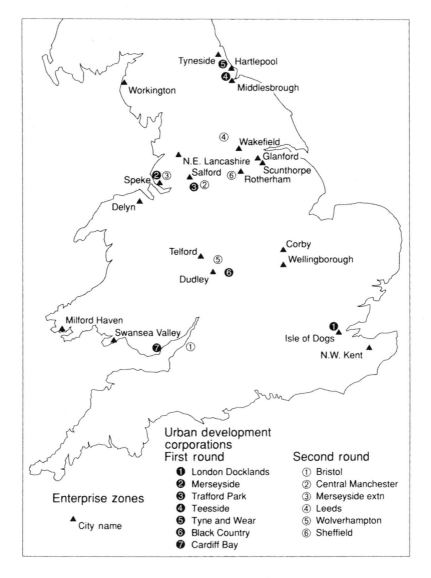

Figure 6.13 Inner-city programmes in Britain, 1988. The enterprise zones were located almost exclusively in derelict industrial or port areas with few residents, though not always in inner cities. The urban development corporations tended to be located in the same kinds of areas, but with a stronger emphasis on the inner parts of the conurbations.

leisure facilities in well designed settings, but the problem has been that other influences have been at work. These include the strong revival of the British economy in the mid-1980s and again in the mid-1990s to mid-2000s, the specific effect of the deregulation of financial markets on the City of London, the growth of one- or two-person households seeking a cosmopolitan inner-city lifestyle (the 'yuppification' phenomenon), the growth of the owner-occupied housing market and the attempts of central and local government to squeeze more new housing into the cities. It seems clear that the rate of inner-urban population and job loss slowed somewhat in the late 1970s and even reversed in some cities (especially London) in the mid-1980s, but that rates of unemployment and related indices of deprivation, in the worst-hit areas, may have spectacularly increased.

What this seems to mean is a new geography of Britain. The old regional dichotomy – between metropolitan south and east on the one hand, peripheral west and north on the other – is being replaced by a more subtle distinction. On one side are the formerly decaying conurbation cities now gradually undergoing transformation into service sector locations, replacing the mining and manufacturing towns in older industrial areas. On the other are pockets of existing middle-class life in the cities that are becoming, in turn, even wealthier, containing the still prosperous suburbs, the medium-sized freestanding cities and towns, and the rural areas of much of Britain. As in the United States, the trends remain strongly towards decentralization of people and jobs, out of the great cities and the older industrial regions, into the environmentally preferable semi-rural areas. Though temporary reversals may occur, as in London during the 1980s, the long-term tendency appears inexorable and irreversible.

Meanwhile, the discovery of the inner-city problem in the late 1970s had direct and immediate repercussions on other parts of urban policy. The most important was that the new towns programme was reduced, while the parallel expanded towns programme – in any case close to completion – was to all intents and purposes phased out. Milton Keynes, the largest and most ambitious of all the Mark 2 new towns of the 1960s, had its target population cut from 250,000 to 200,000: the resources were transferred fairly directly into the revival of the London inner-city partnership areas. However, there was one field where the effects were more muted. In 1976, the 1970-based *Strategic Plan for the South East* was revised by a joint central–local government team, similar to the group responsible for the original plan: in 1978 and again in 1980, the government responded to this revision. While local authority interests were very concerned to reduce the commitment to major growth centres outside London – the Greater London Council because it wanted the resources diverted to London, the Home Counties because growth was politically unpopular with their electors – the government stood firm on the principle of growth in principle, albeit with some studious vagueness about details.

This reflected a striking demographic fact: though estimates of future regional population growth had shrunk with the falling birth rate, migration out of London into the rest of the South East was running at a rate far higher than had been forecast in 1970. Thus estimates of future population growth for the growth areas, and the area around London generally, were reduced, but only marginally. People were apparently voting with their feet, and in a direction contrary to the new policy of inner-city revival. It remained to be seen whether this policy could eventually slow the out-movement, let alone reverse it. Policy, as ever, had to reflect the facts of social and economic change.

The 1980s: policies under Thatcher

During the 1980s the thrust of Thatcherite planning policy was strongly anti-interventionist. No more major regional policy studies were begun, and the tone was set in 1980 by a regional policy statement for the South East on three and a half pages of A4 paper. This and subsequent statements strongly emphasized the theme of unleashing private initiative. Power was progressively pushed down from the counties to the districts, which became the real arbiters of what would happen in the rural shire counties. Within the major conurbations, the Greater London Council and the provincial Metropolitan County Councils – both, ironically, creations of Conservative governments – were summarily abolished in 1986; such strategic planning authorities, the government said in a White Paper presaging the move, represented an outmoded fashion from the 1960s which Britain no longer needed. There was a clear indication that the private housebuilders would be given their head. In 1985 ten of them banded together to form Consortium Developments, which soon announced a major plan to develop new communities – in effect, privately built mini new towns – in the countryside, especially in the South East.

In practice, it did not work out as intended. For the mood of the existing local communities was strongly anti-growth. NIMBY (Not in My Back Yard), a term imported from America, entered the English language. Most of the proposals for new communities were rejected on appeal after lengthy planning inquiries; perhaps the most controversial, Foxley Wood in Hampshire, part of the major growth zone in the old 1970 *Strategic Plan*, was accepted in principle by Secretary of State for the Environment Nicholas Ridley in early 1989 but rejected by his successor Chris Patten a few months later. The crisis of housing in the South East reached a climax in 1987, with press reports that prices in London were rising by £53 a day, and with angry arguments about the calculation of housing needs flung to and fro between the Standing Conference on London and South East Regional Planning, SERPLAN, and the House Builders Federation. What became clear by the end of the decade was that all this was a highly political matter, and that the government was becoming ever more solicitous of its existing shire county voters. The volume builders did not get the freedom they expected. Local authorities, and the government on appeal, were more inclined to allow new industrial and commercial developments which created jobs; so the English landscape was transformed by major edge-of-town industrial estates, warehousing units, hotels and superstores.

But there was not an equal transformation in housing. The reverse, in fact: ironically, this concern prompted the Conservative government, like its Labour predecessor, to a continued stress on inner city regeneration. However, though the general objective remained the same – to target help on the inner-city areas worst-hit by deindustrialization and by the loss of traditional port and other jobs – the chosen instruments were very different.

The enterprise zones

Already in her first 1979 Budget, Thatcher had introduced the principle of enterprise zones: areas of the country which were to be free of normal planning controls, and in which firms were to enjoy a ten-year freedom from local rates (property taxes) and certain other fiscal concessions. During 1980–1, 11 enterprise zones were designated: Clydebank, Belfast, Swansea, Corby, Dudley, Speke, Salford/Trafford, Wakefield, Hartlepool, Tyneside and the Isle of Dogs. They varied in character, from inner cities (Isle of Dogs, Belfast, Salford), through peripheral conurbation areas (Speke, Clydebank) and areas of industrial dereliction (Dudley, Salford, Swansea) to planned industrial areas with services in place (Team Valley in Gateshead); most were blighted urban areas with substantial areas of derelict or abandoned land. A further 13 zones were designated in

1983–4: Allerdale, Glanford, Middlesbrough, north-east Lancashire, north-west Kent, Rotherham, Scunthorpe, Telford, Wellingborough, Delyn, Milford Haven, Invergordon and Tayside. They were more varied in character.

The government introduced elaborate provisions to monitor the success of the zones, culminating in a major independent consultant's review in 1987. It found that from 1981–2 to 1985–6, the zones had cost the public some £297 million net of infrastructure costs which would probably have been incurred anyway; 51 per cent represented capital allowances, 28 per cent rate relief and 21 per cent infrastructure and land acquisition. The result by 1986 was just over 2,800 firms in all the zones, about 70 per cent of them in the ten original first-round zones (excluding Belfast); most were small, only one quarter were new start-ups and most were local transfers. They employed some 63,300 people, but only about half of these could directly be ascribed to the existence of the zone; taking account of all the direct and indirect effects, the consultants concluded, total net job creation in the zone and the surrounding area totalled only 13,000. The cost to the public purse of each additional job created in the enterprise zones themselves was some £8,500; in the local area it was three of four times as much. Firms judged exemption from rates as by far the most important incentive; capital allowances and the relaxed planning regime were also cited.

Overall, then, the enterprise zone experiment produced relatively small numbers of really new jobs, and at appreciable – but perhaps acceptable – cost. Perhaps this is why the third Thatcher government did not extend the experiment; after 1987, it placed far greater weight on the urban development corporation as a mechanism for rapid assembly, development and disposal of urban land; and on the simplified planning regime as a general tool of development throughout the country.

The urban development corporations

The other distinctive device in the Thatcherite alternative policy, curiously enough, borrowed from the device used by the Attlee Labour government of 1945–50 in building the new towns (already described in Chapter 4). This was a public development corporation, financed by the Treasury, and able to exercise powers of land development (including compulsory purchase) in its own right. The urban development corporations (UDCs) assembled sites, reclaimed and serviced derelict land and provided land for development; they could also provide the necessary infrastructure for development, especially roads, and could improve the local environment.

The first two, for London Docklands and Merseyside, were set up in 1981. LDDC (London Docklands Development Corporation), with derelict land only a few miles from the Bank of England, enjoyed a huge success, using £385 million of public money to leverage some £3,000 million of investment commitments; by 1990 it had largely completed the redevelopment of two key areas, the Isle of Dogs (including one of the more successful enterprise zones) and the Surrey Docks on the opposite side of the Thames. Its most conspicuous showpiece was the enormous Canary Wharf development for some 46,000 office workers, including London's highest office tower, in the centre of the Isle of Dogs enterprise zone. Developed by the Canadian developers Olympia and York, Canary Wharf (Plates 6.6a, b) is interesting for the way in which private and public money combined to build the necessary transport infrastructure: an extension of the Docklands Light Railway, and a much more ambitious extension of the Jubilee tube line (Plate 6.7). In the great property recession of 1992 Olympia and York collapsed financially and Canary Wharf was taken into administration by a consortium of banks; but it emerged successfully, to be completed and massively extended a decade later

Plate 6.6 The redevelopment of London Docklands: (a) housing at Wapping: new commercially developed housing for owner-occupation on the site of the old London Docks basin; (b) Canary Wharf: the largest office development in Europe, which provides space for 100,000 workers, seen across the Thames from North Greenwich.

Plate 6.7 Jubilee Line Canary Wharf station. Integral to the success of the regeneration of London Docklands was the provision of new infrastructure. The London Underground Jubilee Line extension from Westminster runs through the Docklands area before linking into a new international station on the Channel Tunnel high-speed line and the London 2012 Olympics site at Stratford. The line was opened in 1999 and each station on the extension was designed by an internationally renowned architect. Canary Wharf was designed by Norman Foster and the station's construction was partly financed by the private sector office complex above it.

after the LDDC had been wound up in 1998. Indeed, with almost 100,000 workers, Canary Wharf now seriously challenges the traditional City of London as a centre for London's financial services.

The LDDC had been successful at redeveloping former derelict areas and securing new job creation. It had even assisted in creating a new business, London City Airport. During the 1990s it encouraged the development of attractive waterside apartments along the River Thames and the former dockside, bringing middle-class residents into the area, to support new shops, restaurants and bars, and physically transform the area in a relatively short space of time.

These original two UDCs were joined in 1987–8 by a whole series of further corporations in English provincial cities – in the Black Country, Teesside, Tyne and Wear, Bristol, Leeds, Manchester and Sheffield – as well as a major exercise in Cardiff Bay. All were time-limited, and had effectively been wound up by the end of the 1990s. Like the enterprise zones, the UDCs concerned themselves largely with areas of derelict industrial or transport land, generally close to city centres. Because of their location, and because the UDCs had generous grants to start the redevelopment process, they did achieve very substantial regeneration, as any visit to a British city will testify (Plates 6.8a, b). But the actual impact in terms of job creation, according to a study made for the government in 1987, was modest: of 63,000 new jobs in the zones, only about 35,000 were a direct result of enterprise zone policy, and most were local transfers; so true net job creation in and around the zones totalled just under 13,000, each of which had cost between £23,000 and £30,000 of public money.

Further, in some cities local authorities achieved spectacular regeneration results without UDC aid – notably in Salford Quays, where the city itself developed the old Ship Canal docks in competition with a UDC in the neighbouring borough of Trafford

(a)

Plate 6.8 The transformation of British cities. The urban development corporations brought about the transformation of derelict areas of older industrial cities by funding infrastructure provision and improving the physical environment, making areas more attractive to market investment and allowing further planning for new developments in the years after the UDCs were wound up; (a) the Gateshead Millennium Bridge linking Newcastle and Gateshead, Tyne and Wear, located in an area already subject to waterside regeneration; (b) the city of Manchester showing the Bridgewater Hall concert venue, the GMEX exhibition area converted from a former railway station, and the landmark 47-storey Beetham Tower skyscraper, opened in 2007.

on the other side of the water, crowning its efforts with the opening in 1999–2000 of two great cultural artefacts: the Lowry and the Imperial War Museum North (Plate 6.9). And in the late 1990s, as such city centre fringe sites have become immensely attractive, city after city has successfully emulated the Salford example.

Other initiatives

The UDCs, like ordinary local authorities, are also able to draw on other related government policies, such as the urban development grant and the urban regeneration grant, latterly combined in the city grant. In addition, inner-city local authorities continued to enjoy a programme which represents rare political continuity: set up by Labour in 1978, it continued under the Conservatives. This was the Urban Programme, which by the late 1980s was mainly being used for projects aiming to strengthen the local economy. But even by the late 1980s the programme was supporting some 90,000 such jobs or training places.

(b)

Plate 6.8 continued

Inner-city housing

There was a major shift in housing policies: as the Thatcher government shrank the public housing new construction programme virtually to nothing, so it could release sites for the private developers. That is why, during the whole of the 1980s, successive Thatcher governments placed such emphasis on monitoring releases of publicly – especially, local authority-owned – land. Helped massively by the existence of the London Docklands Development Corporation, the government was able to achieve its target of getting up to one third of all the new housing starts in the South East within the boundaries of London itself.

This presumably had the desired political effect of creating huge enclaves of owner-occupied housing within what had been safe Labour boroughs. But since it did use land which was otherwise being wasted, it was difficult to argue with it – particularly since the right to buy had already begun to shrink the public housing sector under the previous Labour government. As already seen, it did have the effect of reversing London's long-term population loss. And at the same time, a series of mega-developments arising through new forms of cooperation between public authorities and private developers – especially in London Docklands and around certain key terminal railway stations – brought a net increase in London jobs. But this was not an effect that could continue

Plate 6.9 Salford Quays, The Lowry. An international-quality concert venue and art gallery that opened in 1999 as one of the jewels of the local-authority-led regeneration of Salford Quays, on the former Manchester Ship Canal. The area is also the venue of Imperial War Museum North, a new media centre for the BBC, hotels, commercial venues and apartments.

for long; the necessary land supply was finite. Beyond some point, so long as household formation continued, it seemed certain that the outward decentralization pressures would continue to assert themselves.

Planning in the 1990s: the quest for an urban renaissance

The great reversal continued throughout the 1990s, and London was now joined by other cities which experienced population growth after a half-century of decline. But London's recovery was much more marked than that of the provincial cities, which gained population much more slowly and continued to lose employment as London recorded a modest gain (Table 6.1). Underlying these overall figures were significant migration movements – continuing out of the cities into the surrounding shire counties, but balanced in London's case by strong in-migration from abroad and by natural growth of a young population, less evident in the provincial cities. However, maps produced by the government's Social Exclusion Unit showed that everywhere, including London, the cities were still marked by strong geographical concentrations of multiple deprivation, invariably concentrated in those sections that had suffered the most serious employment losses through de-industrialization in the 1970s and 1980s (Figure 6.14). Though the city centres were thriving through new jobs in the producer and consumer service sectors, and new apartment construction was everywhere in evidence (Plate 6.10a), only a short distance away there were scenes of physical devastation – including, in some cases, housing abandoned and boarded up as the former occupants fled from a mounting spiral of problems – vandalism, crime, drugs, arson – which seemed to be plaguing some areas of some cities (Plate 6.10b).

To these problems, successive governments took rather different views, dependent on their political ideology – and also, to some degree, on the prescription that happened to be fashionable at the time. The first generation of urban development corporations was followed at the end of the 1980s by another group, generally smaller in geographical extent, in many of the major provincial cities. Working particularly around the city centres, they achieved a great deal of physical regeneration in a relatively short time – but, to do it credit, so did a city like Salford which regenerated its waterfront itself. London Docklands was followed in March 1991 by a much bigger scheme: Thames Gateway (Figure 6.15): a corridor more than 30 miles (50 kilometres) long, stretching through East London and the neighbouring parts of Essex and Kent, following the line of the planned high-speed rail link from London to the Channel Tunnel, and with major urban regeneration sites around the stations and in intermediate locations. The subject of a major planning exercise since the mid-1990s, Thames Gateway has been slow to come to reality because of a variety of factors, not least a delay in starting construction of the new rail link.

One other reason was that, unlike Docklands, there was no specific agency charged with the task of regeneration and primed with government money to start to achieve it. In part this was because the responsible minister at the time, Michael Heseltine, based his plans on the creation of a super-development corporation spanning all of England and modelled on the successful Scottish Development Agency. Finally born in 1994, English Partnerships had a somewhat mixed career in the following years, partly because of its association with the controversial Millennium Dome at Greenwich.

Table 6.1 *British cities: population and employment change*

Population change, 1951–99

	1951–71 Change (%)	1971–91 Change (%)	1991–9 Change (%)
London	−9.5	−14.2	+13.9
Birmingham	−1.3	−14.6	+8.0
Liverpool	−22.7	−26.2	+1.8
Manchester	−22.6	−26.5	+7.8

Employment change, 1981–96

	1981–91	1991–96
London	−8.6	+2.9
Birmingham	−8.5	−3.4
Liverpool	−23.2	−12.3
Manchester	−11.4	−9.4

Source: Decennial censuses; *Population Trends*; I. Turok and N. Edge (1999) *The Jobs Gap in Britain's Cities: Employment Loss and Labour Market Consequences* (Bristol: Policy Press)

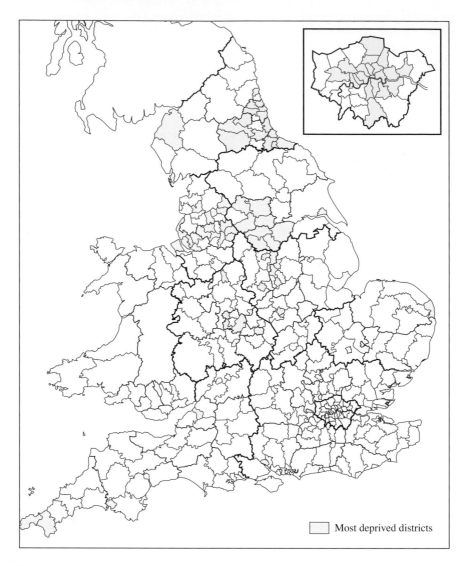

Figure 6.14 Multiple deprivation in the UK, 1998. The first report of the Social Exclusion Unit shows how exclusion is highly concentrated in the major conurbations, and even in certain wards within them.

At this point, much of its work was effectively subsumed in that of the Regional Development Agencies (RDAs), established by the Blair government of 1997 as a new means of regional regeneration.

During the 1990s the favoured tool of urban regeneration was the Single Regeneration Budget (SRB). Developed out of an earlier experiment, City Challenge, the SRB represented the strong belief of Conservative ministers in the virtues of competition. Cities and towns were encouraged to submit competitive bids each year for imaginative schemes which were funded from a single source covering all aspects of regeneration. The scheme generated a great deal of enthusiasm and hard work, but also a great deal of disappointment, and critics argued that the distribution of funds reflected grantsmanship more than the intensity of problems in an area. Responding to this charge, the second

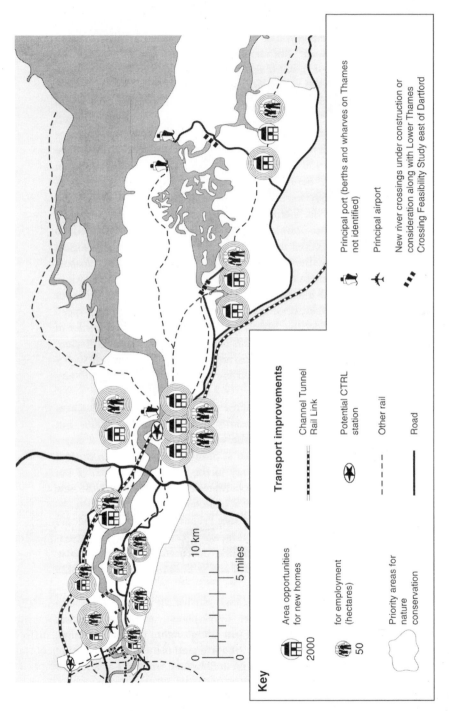

Key

Area opportunities
for new homes

2000

for employment
(hectares)

50

Priority areas for
nature conservation

Transport improvements

▬▬ Channel Tunnel
Rail Link

✳ Potential CTRL
station

- - - Other rail

▬▬ Road

🚢 Principal port (berths and wharves on Thames
not identified)

✈ Principal airport

▰▰ New river crossings under construction or
consideration along with Lower Thames
Crossing Feasibility Study east of Dartford

10 km

5 miles

0 — 5 — 10

Figure 16.5 Thames Gateway: Regional Planning Guidance 1995. This has become the most ambitious urban regeneration scheme in Europe: a 30-mile (50-kilometre) discontinuous corridor of development along the lower Thames, based on the high-speed Channel Tunnel Rail Link.

(a)

Plate 6.10 A tale of two housing markets. (a) New apartments, the Norman Foster-designed Albion Wharf complex at Battersea, south-west London, built on a brownfield site. In 2010, prices to purchase apartments here ranged from £500,000 to £4 million. (b) Deptford in south-east London, less than 1 kilometre from Canary Wharf financial centre and Greenwich on the south side of the Thames, an area that has suffered economically with the closure of the dockyard and still has high concentrations of deprivation and unemployment. In 2010 the average residential property price in this deprived inner-city area was £280,000, still over ten times the average London salary.

Blair government of 2001 dropped the scheme and transferred the funds to the RDAs to distribute them as they determined. By this time, however, the stress was even more on 'joined-up thinking': under the New Deal for Communities programme, different agencies within an area – whether part of the local authority or not – were encouraged to collaborate to produce coordinated answers to problems of deprivation, whether these lay in the area of planning or of education or of social service provision.

These broad-based programmes did, however, have another thrust. From 1996 onwards, the Major and then the Blair government were compelled to respond to new household projections which proved politically contentious. They showed that against a background of a modestly increasing population, there were likely to be some 4 million new

(b)

Plate 6.10 continued

households in England over the following quarter century – four fifths of them consisting of a single person living alone, the product of complex socio-demographic changes: more young people leaving home for higher education, more divorces and separations, more widows and widowers surviving their partners for longer and, most intriguing, many more people apparently choosing to live alone. The projections meant a sharp upturn in the forecast needs for new housing, and this was compounded by population increase, partly as a result of continued immigration and high rates of natural increase among the immigrant communities, in London but not so noticeably in other big cities (Table 6.1). The resultant housing pressures proved most contentious in the south of England, where they met fierce NIMBY opposition, locally from electorates, and nationally from the Council for the Protection of Rural England (CPRE). The government's reaction was to defuse the opposition by appointing a high-level Urban Task Force, under the chairmanship of the architect Richard Rogers, to look at the entire complex array of problems. In its report, published in 1999, the task force neatly linked them: urban abandonment and greenfield development, it concluded, were two aspects of the same syndrome. It argued that both could be countered by a campaign to forge an urban renaissance, which would again make British cities attractive places to live, thus reclaiming the abandoned areas and reducing the pressures on the surrounding countryside; it produced design prescriptions for the kind of development it wanted to see in the cities (Figure 6.16); and it offered no less than 105 policy recommendations to help achieve this renaissance.

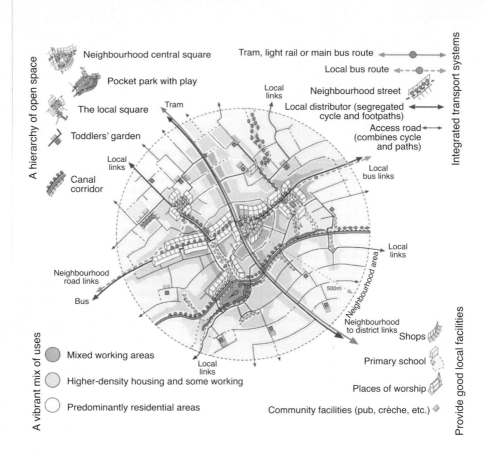

Figure 6.16 Urban design principles, from the Urban Task Force report. Heavily influenced by the ideas of the American 'new urbanism', this diagram represents a return to traditional urban design principles – including a reversal of the precinctual principle developed by Sir Alker Tripp and later embodied in the Buchanan Report.

The government's response was an Urban White Paper, 18 months later; it accepted some but by no means all of the recommendations, some of which would have proved costly to the Treasury. But in any case, the fact was that the problem was rather different in kind in the south and in the north of England. In the south, London was booming and the problem was to find enough brownfield land to build upon; around it, the problem was that too much development was taking place at excessively low densities, in places that made it completely car-dependent. In the north, the problem is that too few people – apart from a minority of loft-living pioneers around the centres – wanted to live in the cities; they were fleeing to the surrounding areas, and here local authorities often proved only too willing to plan the housing to accommodate them (Plate 6.11). True, the two cases did have some things in common: everywhere, it proved difficult and expensive to reclaim brownfield land, because it was often contaminated and had other physical problems; indeed, many of the task force recommendations were directed at removing or reducing that anomaly. And everywhere rural planning authorities were

Plate 6.11 Liverpool waterfront, Prince's Dock. An example of the regeneration of former industrial areas in an inner northern city and the provision of new high-rise apartments targeted at one- and two-person professional households.

inclined to welcome low-density development because it would prove more acceptable to existing residents. But there the similarity ended: the figures in Table 6.1 demonstrate just how different were the situations in London and the great provincial cities.

Meanwhile, the post-Second World War planning system itself was entering some kind of crisis: not terminal, perhaps, but prompting a basic interrogation into the way it was functioning, half a century after it had come into being. In 1991 the government had modified the comprehensive Town and Country Planning Act it had passed only a year earlier – essentially, a consolidation and updating of the original 1947 legislation – to introduce what it called a plan-led system: under Section 54a of the amended 1990 Act, the adopted development plan became the 'principal material consideration' in determining an application to develop, or an appeal against refusal of permission. This was a marked contrast to the approach of the 1980s when the development plan, and indeed local authority powers, had been effectively bypassed or ignored in favour of market-led projects and centrally guided policies aimed at employment creation and urban regeneration. The government hoped that it would introduce an element of certainty and clarity into the system, simplifying the entire process of development control and speeding appeals. In reality, the government's experience of mediating in major house building projects like Foxley Wood between housebuilders and anti-development Conservative shire voters may have prompted them to hand back control, and ultimately blame, to local government. But underlying all this was a sense on the part of business, and underlined on their behalf by the Department of Industry and the Treasury, that the planning system was acting as a brake on entrepreneurialism and reducing the UK's economic competitiveness. In the event, little changed on the ground since other material considerations could still upturn the contents of the development plan while the government continued to emphasize the role of the private sector in leading development projects while shaping the system through a series of national planning policies, the Planning Policy Guidance notes, or Planning Policy Statements as they were later to become. Individual proposals were still subject to intense pro-development/anti-development battles on individual sites between the developers and residents with the planners attempting to referee the process in the middle with a revised development plan process that remained relatively weak.

Two flashpoints came at the start of the new millennium. One was in the South East, where a bitter battle occurred over Regional Planning Guidance. The regional federation of local planning authorities, SERPLAN, proposed a house building target of only 35,000–37,000 units a year; a government-appointed panel conducted an inquiry and substituted 55,000; the government first split the difference at 43,000 and then, with a general election looming, cut this to 39,000. Another came in the Cambridge region, where high-technology growth was creating one of the most dynamic parts of the entire UK economy. The government itself rejected a proposal for a biotechnology-based research and development centre in the green belt to the south of the city, prompting a furious reaction; finally, in 2001, a revised version of Regional Planning Guidance proposed to steer most growth north of the city, while warning that the scale of development would exceed that of Milton Keynes in its years of maximum development.

Into the 2000s: planning under pressure

Upon taking office in 1997, perhaps remarkably, the Labour government possessed no 'big agenda' for the planning system. In fact, planning hardly featured on the government's first-term programme (1997–2001), though the 1999 Urban Task Force report and the following 2001 White Paper could be said to start the process of generating one for a second term. A 1998 statement by the minister for planning discussed improving efficiencies but did not hint at any radical change. If anything, the period can be characterized as business as usual, since the emphasis was placed on aspects of constitutional change and social-policy improvements – core areas of the Labour Party's manifesto pledges. There was an indirect impact on planning, caused by devolution of power to the Celtic countries and London and a commitment to enhance regional governance in England (see Chapter 5), but this was scalar in nature with the emergence of new policy tools and levels of responsibility. Yet three problems nagged away. The first was a growing problem over housing, particularly concerning rising house prices, the lack of affordable housing and the mismatch in supply and demand between the south-east of England and the north of England. The second concerned the speed and perceived inability of the planning system locally to deliver change and developments to the timeframe to satisfy businesses and ministers. The third concerned the competitiveness of UK cities both globally and in relation to the rest of the UK, and the appropriate governance mechanisms to deliver successful growth to benefit the country as a whole – a point brought into sharp relief by the publication of the *State of the English Cities* report, a major research milestone which adduced a huge mass of statistical data demonstrating not only that the economies of London and its surrounding towns were performing much better than their northern equivalents, but – equally significant – that major northern core cities like Manchester and Leeds were performing significantly better than the old manufacturing towns and seaside resorts in their respective regions.

What emerged in the south was an intensifying problem of housing affordability in the first years of the new millennium, as low interest rates combined with rising incomes and higher rates of migration, to generate an escalation of house prices. In the south of England, housing rapidly became unaffordable for a substantial minority of the population (Figure 6.17), even including people on moderate incomes, such as many key workers in the public services. In 2005 property prices in London were ten times the average salary. But despite the pressure for cheaper housing, new house building has been in decline since its peak in the mid-1960s (Figure 6.18). As the sell-off of social housing from the early 1980s was accompanied by an almost complete halt to traditional council-house building, the housing association (registered social landlord) sector and the private-house building industry failed to fill the gap in the need for social housing while the private housebuilders failed to respond to the increased demand for home ownership. The builders pointed to the failure of planning to release adequate amounts of land for new housing development, but there was scant hard evidence to back up this assertion. In fact, a study by the Royal Town Planning Institute indicated significant amounts of land with planning permission for housing remained undeveloped as housebuilders 'landbanked' supply. To economists and particularly the Treasury, the problem lay at the door of planners. Successive government ministers in the Labour government criticized the unresponsiveness of planners to market demands.

This was exacerbated by projections for household growth, and critical examinations of the planning and housing relationship in a report by Kate Barker, a Treasury official, in 2004, suggesting that there would be 155,000 additional households each year on top of an accumulating backlog; as a result, the government first estimated in 2005 that more than 200,000 additional households would require housing each year, and then in 2007 increased that figure to 300,000 a year. As stated earlier in the chapter, many of these will be single-person households, reflecting changing social trends, changing pattern of relationships, higher divorce rates, rising wealth and people living longer. However, successive official reports showed clearly that the nature of the housing problem was quite different in the south and in the north of England – effectively, in fact, opposite. In the South, the problem is shortage of supply at almost every price level, affecting

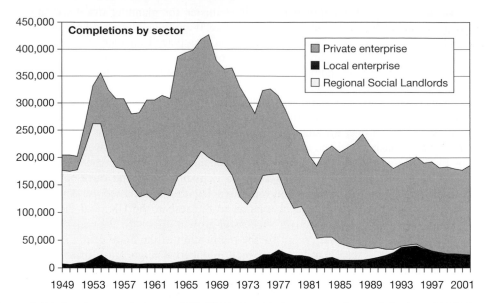

Figure 6.17 Housing completions, 1949–2001. Completions reached a peak in the mid-1960s, as slum clearance and rebuilding propelled a huge programme of public housing simultaneously with buoyant private construction. But in the 1970s, and increasingly throughout the 1980s and 1990s, public construction drastically shrank as local authority building effectively came to an end and housing associations failed to make up the gap. In 2008, when recession brought a slump in private construction, overall completions sank to the lowest levels recorded since the early 1920s.

not merely subsidised housing for low-income households but also market housing for middle-income groups. Prices have risen continuously but at variable levels since the mid-1990s, and many on modest incomes, including key public sector workers and others essential to the local economy, cannot afford to buy, with a further knock-on effect to the rental sector as many purchasers buy-to-let at high rent levels. In sharp contrast, parts of the North, especially old manufacturing towns that have lost their former economic base, have suffered the opposite problem with housing market collapse, leading to homes or even streets being abandoned. Additionally, both in the South and in the North, migration – within regions, from urban to rural – has also become widespread, as people seek a better quality of life in the suburbs and exurbs, in turn exacerbating the potential for further urban sprawl. In the North this trend has been accompanied by large-scale housebuilding on greenfield sites, often on the edges of existing towns and cities. Between 1997 and 2000, for example, 54 per cent of new dwellings were built on greenfield land in the North East, and nearly 50 per cent in Yorkshire and the Humber. All these problems bring more pressing, and rather different, housing needs than the UK has witnessed previously.

There has been a significant rise in owner occupation over the period, partly achieved through rising incomes and the right-to-buy legislation of the 1980s where the social-housing sector was sold off to private buyers. And yet despite the rise of owner-occupation, the private sector has failed to meet the challenge: new house building fell steadily from a peak of 350,000 annually in the late 1960s to below 140,000 in 2003 and then even more catastrophically, as the recession took hold, to less than 80,000 in 2009. The net figure is actually even lower than this, when taking account of demolitions and conversions. Private housebuilding has therefore failed to rise to meet the demand

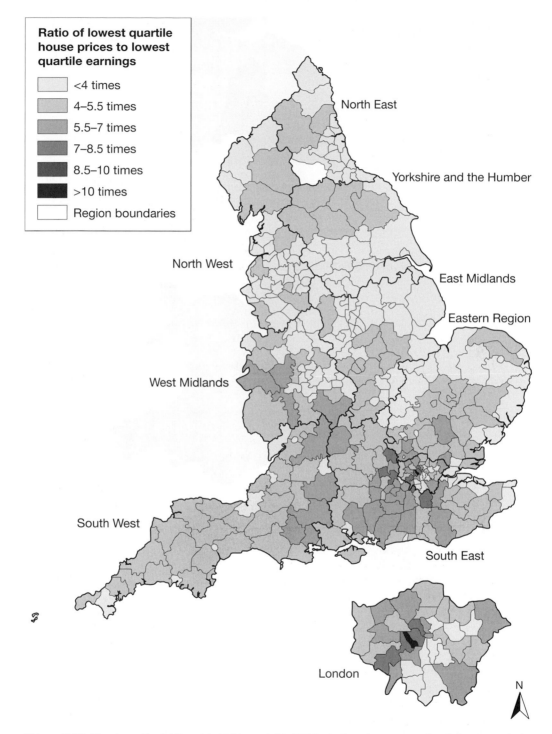

Figure 6.18 Housing affordability: (a) 1996; and (b) 2006. As housing construction fell progressively behind rising demand in the boom years around the millennium, it was inevitable that housing would become less affordable. This problem dramatically spread geographically: restricted to England's south-east corner and parts of the Midlands in 1996, a decade later it had spread to affect much of the country.

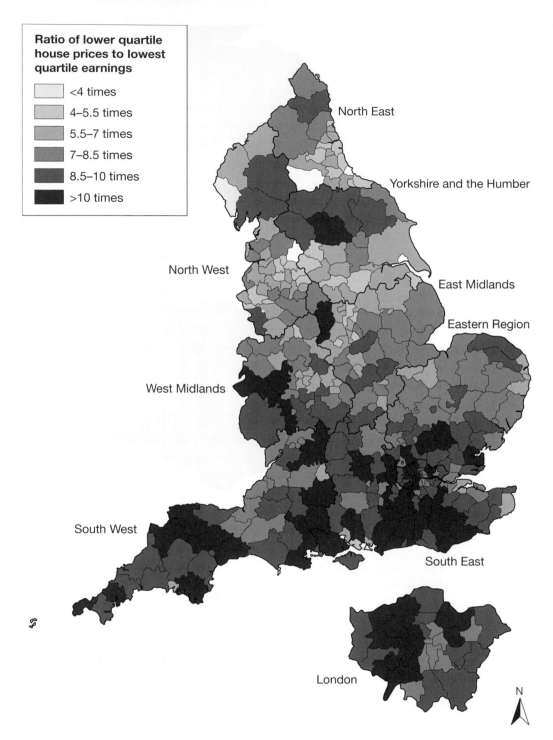

Ratio of lower quartile house prices to lowest quartile earnings

- <4 times
- 4–5.5 times
- 5.5–7 times
- 7–8.5 times
- 8.5–10 times
- >10 times

North East

Yorkshire and the Humber

North West

East Midlands

Eastern Region

West Midlands

South West

South East

London

N

Figure 6.18 (b)

for owner-occupied property, with too many large homes being built when the new demand is mainly for small households, caused by the demographic shift. With the continued dominance of the larger volume housebuilders, employing standard housing designs and larger 'executive' style property ranges, there has been an apparent unwillingness for housebuilders to react to the social change statistics and adapt their own plans accordingly. The national picture masks a more varied picture at the regional level.

The government's eventual response, launched in February 2003 by Deputy Prime Minister John Prescott, was the Sustainable Communities: Building for the Future plan. This set out a long-term programme of action for delivering sustainable communities in both urban and rural areas, by aiming to tackle housing supply issues in the South East, low demand in other parts of the country, and the quality of public spaces. The £22 billion programme encouraged the design of more sustainable forms of social, economic and environmental development through the management and distribution of economic growth, with an emphasis on mixed use, cohesive communities, sustainable living and high-quality design. There were invariably tensions within the programme, not least the strategic requirement to sustain the economic growth of London and the South East on the one hand with the need to generate a more balanced economic growth and competitiveness in other English regions by tackling the problems of low demand and poor-quality housing. In the spirit of spatial planning and coordination, part of the effort of the programme was also to focus attention and coordinate the efforts of all levels of government and stakeholders in bringing about development that meets the economic, social and environmental needs of future generations. National policies are brought together but also given a regional dimension. More controversially, the plan recognizes the need for stock reduction through demolition in the north of England as the only viable alternative.

The plan consists of several key elements. First, there is the requirement to address the housing shortage, particularly by accelerating the provision of housing. This includes ensuring that housing numbers set out in regional planning guidance for the South East can be delivered, by accelerating growth in the four 'growth areas' (Thames Gateway, Ashford, London–Stansted–Cambridge–Peterborough, and Milton Keynes–South Midlands, and by ensuring that the construction industry has the right skills to deliver. There is a recognition that affordable housing needs to be provided in a more robust way. The plan allocated £5 billion for the provision of affordable housing over the next three years, including £1 billion for housing 'key workers' in the public sector, to aid recruitment and retention. Alongside the affordability issue, the plan also aimed to tackle homelessness, by ending the use of bed and breakfast hostels for homeless families. The housing problems of the North are very different to those of the South, and rather than deal with housing supply – a key requirement of the South – the programme for the northern regions tackles low demand and abandonment; here, the plan estimated that there were around 1 million homes in parts of the North and Midlands that might need either demolition or total renewal. Nine Market Renewal Pathfinder schemes were established in the areas worst affected to put in place action programmes to address this problem. These comprised Birmingham and Sandwell; East Lancashire (Blackburn, Hyndburn, Burnley, Pendle, Rossendale); Humberside (Hull and East Riding of Yorkshire); Manchester and Salford; Merseyside (Liverpool, Sefton and Wirral); Newcastle and Gateshead; North Staffordshire (Stoke and Newcastle-under-Lyme); Oldham and Rochdale; and South Yorkshire (Sheffield, Barnsley, Rotherham and Doncaster) (Plate 6.12). In 2004 a separate document was published dealing with the problems of the North, entitled *The Northern Way*. Both the South and North documents offer sub-national strategies to tackle two very different housing markets and economic issues (see Figure 6.19).

Plate 6.12 Moss Side, Manchester. A multi-ethnic area, Moss Side has suffered high levels of deprivation, poor-quality state housing and crime incidents. In the 1980s and 1990s, it was the location of riots and trading in narcotics. Since 2003 the area has improved with redevelopment and regeneration of the housing stock and local environmental schemes and the provision of local public services. Moss Side is part of the £361-million state-funded Manchester Salford Housing Market Renewal Pathfinder Scheme targeting deprived communities and housing market collapse.

The Sustainable Communities programme can be seen as a reassertion of traditionally orthodox British spatial planning principles, in its concern to shape a very ambitious pattern of sustainable spatial development supported by social, economic and environmental resources in the interests of the long term and of future generations. In a statement that bears some resonance to the origins of modern planning and of the establishment of the town and country planning movement over 100 years ago, as well as the new towns programme of the 1940s and 1960s, the plan states that:

> Investing in housing alone, paying no attention to the other needs of communities, risks wasting money – as past experience has shown. A wider vision of strong and sustainable communities is needed to underpin this plan . . . Places where people want to live and will continue to want to live.

This re-awakening of a broader vision for places is interesting, partly because of the link back to the purpose of town planning in the early twentieth century, but also because it has become a distinctly twenty-first-century objective of sustainable economic competitiveness of cities and regions, as drivers of economic growth in the global economy. It is telling that in the parlance of twenty-first-century government too, the

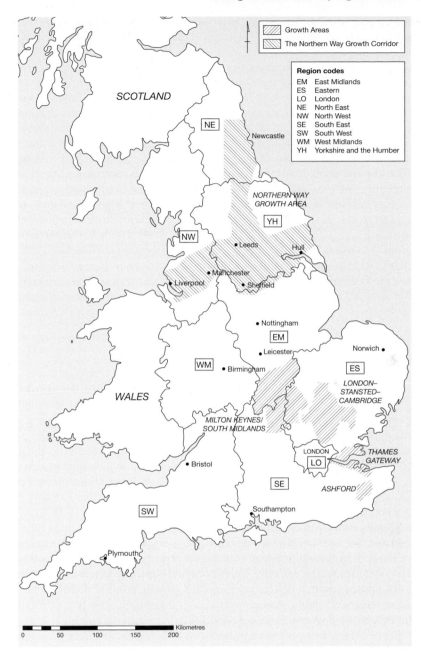

Figure 6.19 The Sustainable Communities Growth Areas of 2003 and the Northern Way of 2004. The Sustainable Communities growth areas were designated by the deputy prime minister in 2003 and cover Ashford, Milton Keynes and Northamptonshire, Stansted–Cambridge and the Thames Gateway. Intended to channel and accommodate housing growth in the Greater South East of England, the growth areas extend London's influence over a large territory but utilize different delivery vehicles to deliver development. The Northern Way strategy, Moving Forward, was launched in September 2004 and covers the three standard northern regions of the North West, the North East and Yorkshire and Humberside. The strategy concentrates on key investment priorities, collaborative working across agencies and securing delivery through – uniquely in governmental terms – city regions of eight conurbations. The intention has been to tackle the £30 billion gap in output between the North and the average for England.

attention is not towards 'planning' but towards 'sustainable communities', even though some of the broader objectives of this step change are directly concerned with the very existence and purpose of planning.

Further government intervention in the strategic housing market occurred in 2007 when it was announced by the Department for Communities and Local Government that a competition would be held for the development of ten eco-towns across England. These eco-towns have proved to be controversial even though the original intention behind the idea was to achieve high standards of sustainable living while also maximizing the potential for affordable housing. At least one third of the new housing is intended to be allocated as affordable. Some of the towns are planned to accommodate 20,000 new homes, and should be 'zero-carbon' developments and exemplary in one area of sustainability, such as energy production. The new environmentally friendly towns – low-energy, carbon-neutral developments built from recycled materials – are intended to be largely car-free, with pedestrian and cycle-friendly environments. Following an initial shortlist of 17 sites that were identified in 2008, the government announced the first four eco-towns in July 2009, at Whitehill-Bordon in Hampshire, St Austell in Cornwall, Rackheath in Norfolk and north-west Bicester in Oxfordshire. Some £60 million has been made available from the public sector to fund infrastructure provision. Further sites are proposed after 2010 but are dependent on the policies of the Coalition government.

Planning reform, planning gain and localism

Another characteristic of urban and regional planning in the first decade of the twenty-first century was structural reform. Most noticeably, December 2001 saw the publication of a planning Green Paper that discussed major reform of the tools of planning, particularly in relation to development planning. Terms such as 'fundamental change' and 'radical overhaul' were used by ministers to herald the Green Paper. Furthermore, the government seemed unsure exactly what to do with planning as a central government area of responsibility, with the host department changing titles from Department of the Environment (DoE), to the Department of the Environment, Transport and the Regions (DETR), to the Department for Transport, Local Government and the Regions (DTLR), to the Office of the Deputy Prime Minister (ODPM) in a mere five years, and again to the Department of Communities and Local Government (CLG) in ten years, all with slight changes of remit. These departmental changes were not necessarily solely related to how government accommodated planning centrally, but may nonetheless reflect tensions within a political party committed towards delivering economic growth, environmental protection, social inclusion, enhanced public transport provision, decentralization and devolution.

Mindful of the ongoing criticisms, and initiating legislative change promised by the 2001 Green Paper, the government attempted to bring about direct planning reform to address some of the perceived inadequacies of the tools of planning. This was intended not necessarily to reinstate a corporate role for planning, but rather to make planning more efficient for development implementation by the private sector and intended to modernize planning as a public service for the twenty-first century. In contrast to the reforms of the 1980s and 1990s, when the intention was to focus planning on the regulation of negative externalities, reforms under the Labour government have – perhaps somewhat surprisingly given the lack of discussion about planning in the first term in office – potentially reinvigorated planning to become a proactive coordinating activity, intended to assist in delivering development as part of continued economic growth. A hallmark of this transition has been to commodify planning – performance league tables, assessing the quality and quantity of decisions, and the desirability of the process in meeting short-term needs – to the point where speed and efficiency are features that now dominate the planning system.

The most significant element of planning reform in this process was the Planning and Compulsory Purchase Act 2004, which completely reformed the planning policy and strategy-making function at national, regional, sub-regional and local levels. From 2004 the revised planning policy framework comprised National Planning Policy Statements that replaced Planning Policy Guidance Notes; Regional Spatial Strategies replacing Regional Planning Guidance Notes; Sub-Regional Strategies, introduced for the first time; and Local Development Frameworks, Area Action Plans and Masterplans replacing the old hierarchy of Structure Plans, Local Plans and Unitary Development Plans. And a key objective of the planning reforms is to 'front-load' the planning system by introducing more opportunities for public consultation within planning policy-making. This is how the government has redefined spatial planning as it applies to the British planning systems. The reforms do not merely change the titles of the tools of planning, but also broaden out their role in public sector service delivery and still encompass land use planning; essentially the new planning documents become, at each level of administration, the 'plan of plans'. After 2004 the British planning system can be characterized as comprising two very different styles of planning: traditional land use planning process; and spatial planning that is more concerned with strategy and actor integration and delivery of place-based mechanisms. Although the reforms were well intentioned, it soon became apparent that the change was taking considerable time to initiate practically within local government and regional governance structures. The new tools took just as long to prepare as the old documents; if anything, the expectations on front-loading the system to address the evidence of a range of social, economic and environmental issues locally, coupled with planners' uncertainty in how to progress the new requirements, resulted in a number of Local Development Frameworks being found to be 'unsound', and in understandable political impatience on the part of ministers. Perhaps little had been learned from the 1950s: heighten the plan's strategic long-term role and it jars with addressing short-term issues.

With the 2004 legislative reforms still occurring in practice, the government sought to place a new emphasis upon delivery. Two major reports undertaken by the Treasury official Kate Barker in 2004 and 2006 sought to address the relationship between housing supply and planning, and economic growth and planning respectively. These provided relatively well balanced reviews of the role of planning regulation in managing and delivering housing and job creation opportunities and coincided with the Sustainable Communities Plan and the 2004 Planning Act. But it was clear that deregulation and the faster delivery of projects were the core government themes driving forward policy, even if the new planning tools were going to take some time (perhaps too long) to enact. Housing affordability, economic competitiveness and climate change provided the direction and underpinned two new planning bills (subsequently to become the Planning Act 2008 and the Climate Change Act 2008) and a range of secondary legislative and policy changes that changed the landscape and practices of planning further. As we discussed in Chapter 5, a key theme of these reforms concerned the national level, with the introduction in 2010 of a new Infrastructure Planning Commission, a centrally appointed committee charged with taking over responsibility for determining major nationally significant development schemes such as airports and power stations from local government, and indeed free from the need for ministerial approval – a politically contentious point, resisted by the Conservative opposition. Further elements of the reforms permitted deregulation of planning controls over small-scale developments. One key theme of the change is the extent to which planning will no longer be a 'passive'

Plate 6.13 British city urban renaissance. (a) The Liverpool One complex, a shopping, leisure and residential development of 42 acres (17 hectares) and 160 stores in Liverpool city centre opened in 2008, the largest city centre redevelopment project in Europe since the postwar period. Significant attention is paid to the quality of the architecture and the public realm. (b) Bristol's Cabot Circus development, also from 2008, comprising multi-level pedestrian streets and walkways, and over 150 shops and services in a mixed-use development.

actor in development delivery. Housing delivery targets will be financially incentivized, carbon performance will be measured and targets for economic growth will be set, requiring a more active and interventionist model of governance and planning. It will take some years to assess the degree to which these planning legislative reforms do actually deliver developments in a more efficient and more economic way. In the meantime, British towns and cities have been subject to comprehensive redevelopment led by the market with a concentration on improving urban design and the urban public realm (Plate 6.13a, b).

A further element of planning reform, introduced in the Planning Act 2008, related to the finance of infrastructure and planning gain. The role of planning obligations in the planning process has evolved through time, from the original concept of betterment, first introduced under the 1947 Act, through to modern day Section 106 (S106) agreements under the Town and Country Planning Act 1990. Throughout this time, the purpose of planning gain has shifted, from the initial attempts to extract money from developers through betterment and to compensate those losing out, through to modern day planning gain, which has traditionally sought to mitigate the impacts of development through S106 agreements. Local authorities are increasingly using S106 agreements to secure the infrastructure needs arising under the government's sustainable communities growth

Plate 6.13 (b)

agenda. Mechanisms are currently sought to help secure developer funding towards these more regional and sub-regional infrastructure needs.

Following the introduction of the Planning and Compulsory Purchase Act 2004, the government considered new approaches to planning gain, with a new agenda of delivering not only site-specific infrastructure and services but also delivery of wider sub-regional facilities underpinning the growth agenda. The need to revise the current system of obligations has long been recognized. In 2001 the government published its paper *Planning Obligations: Delivering a Fundamental Change*, which put forward plans to revise the use of planning obligations, which had been criticized as being inconsistent, unfair and lacking in transparency and a source of substantial delays.

The debate instigated by this paper led to the publication of Circular 05/2005 and inclusion in the Planning and Compulsory Purchase Act 2004 of amendments to S106 powers. Two options were put forward: a Planning Gain Supplement (PGS) and an Optional Planning Charge (also known as tariffs). The PGS represented a modern policy stance on the concept of betterment, which has been implemented on numerous occasions by previous Labour governments. The government's proposals were, according to the Treasury in 2005, to use PGS 'to help finance the infrastructure needed to stimulate and service proposed housing growth, and ensure that local communities better share in the benefits that housing growth brings'. PGS was effectively a tax on development value, collected by the Treasury, but proposed to return the majority of revenues to the region where they arise, and thus contribute to key strategic regional and local infrastructure. By contrast, the Optional Planning Charge (OPC) would have taken the form of a standard tariff on any new development. One example of where this has been implemented has been Milton Keynes, where the so-called 'Milton Keynes roof-tax' has been implemented at a fixed price of £18,500 per dwelling to take into account the likely costs of any development on the environment and local community.

After much procrastination, the government implemented its proposals in the Planning Act 2008 but shied away from the radical reform of planning gain that it had contemplated in 2001. Rather, a new process, the Community Infrastructure Levy (CIL), was introduced. Under this scheme, a proportion of the increase in value on land as a result of planning permission is used to finance the supporting infrastructure, such as schools, and will 'unlock housing growth'. Local authorities are now preparing Local Infrastructure Plans to amass and analyse the local infrastructure needs and to translate these into a programme to allocate proceeds of the levy. But this is predicated on development permission being given and implemented, and the finance being raised to allocate to spend on infrastructure; at a time of economic recession, with less development occurring, there is a danger that these plans remain as wish-lists for the moment. Meanwhile the Conservative opposition, in their Planning Green Paper of 2010, undertook that if elected they would substitute a tariff linked to provision of local infrastructure.

The middle of the decade was also marked by a concentration in policy terms on urban and neighbourhood regeneration, urban and regional growth and capacity, and the creation of non-directly elected governmental bodies at the local level alongside local government to tackle policy delivery and implementation. This 'localism' programme of the Blair government was different from previous approaches for urban renewal and planning. A Neighbourhood Renewal Fund (NRF) was formulated, intended to be a comprehensive approach to tackle the problems of neighbourhood decline. These problems focused around local areas were identified as high unemployment and crime rates, economic change, declining old industry, the demands of new skills, poor housing and the physical environment. Interestingly, and unlike previous governments' initiatives, the response seems a departure from property-led regeneration policies, although the involvement of non-local government actors in the new responses may have marked a continued search on the part of Whitehall to find a range of delivery mechanisms, other than local government, to implement community policies. Business Improvement Districts (BIDs), transplanted from the United States, were introduced in urban areas as public–private partnerships in which businesses in the area paid an additional tax in order to fund improvements and manage those assets within the area's boundaries. Local Strategic Partnerships (LSPs) were also established in 88 NRF areas. These brought together at a local level the different parts of the public sector as well as private, business, community and voluntary sectors to enable different initiatives and services to work together. They were a requirement for NRF funding eligibility funding initially but were eventually rolled out more widely after 2005.

LSPs were tasked with developing and delivering a local neighbourhood renewal strategy to secure more jobs, better education, improved health, reduced crime, and better housing, narrowing the gap between deprived neighbourhoods and the rest, and contributing to the national targets to tackle deprivation. Community strategies were to be overarching documents which brought together development plans and other plans and strategies, partnerships and initiatives to provide a forum through which mainstream service providers (local authorities, the police, health services, central government agencies and bodies outside the public sector) worked together to meet local needs and priorities.

For planning, this smacked of both familiarity and change. The themes of both the LSP and the community strategy were reminiscent of some of the tasks imbued (not always successfully) within planning – essentially corporate planning – under the welfare state before 1979. The differences today meant that planning was not given a central corporate task to coordinate responses, partly because of the decline of the welfare state but also because of increased scepticism on whether planning as a discipline and a profession could actually deliver change. Planning's role in the twenty-first century required working with a non-statutory decision-making body that might set objectives and priorities for a development plan. It was, in effect, a move broadening out the process of spatial planning to a range of planning and non-planning organizations. The task of policy planning began to diverge sharply between urban and rural areas; the process of managing the built environment and delivering public services in the cities became less controlled by planning and more led by a range of governance actors, with the planning process providing a supporting or facilitating role in enacting change through the partnership ethos. Only this type of approach would begin to address the multifaceted problems facing places in the twenty-first century. The political attack on silo thinking was not confined to planning, but lifting the focus of planners towards an integrative strategy required new skills and attitudes. For some in the planning profession, this transition has been too radical to contemplate.

Meanwhile, a parallel problem emerged: the appropriate spatial units to deliver the new hierarchy of plans. Regional spatial strategies, in particular, logically demanded a new regional unit, but this suggested a transition to regional government following devolution to Scotland and Wales, where the new governments both produced spatial strategies. John Prescott's ill-fated attempt to offer regional governance to the North East region, defeated in the referendum of 2004, eventually resulted in a compromise that satisfied no one, whereby the unelected Regional Development Agencies were made responsible for producing regional spatial strategies. But meanwhile government thinking was evolving in a different direction: sub-regional plans for larger metropolitan areas, produced by voluntary agreement among unitary authorities. The first of these, for Greater Manchester, took formal shape in 2010.

Further reading

Useful sources here are Cullingworth and Nadin (2000), Cullingworth (ed.) (1999) and Cherry (1996) (see Further reading, Chapter 4). For a fuller treatment of urban growth, see Peter Hall, Ray Thomas, Harry Gracey and Roy Drewett, *The Containment of Urban England* (Allen & Unwin, 1973).

For the inner-city problem, see Peter Hall (ed.), *The Inner City in Context* (Heinemann, 1981); and Department of the Environment, *Inner Area Studies: Liverpool, Birmingham and Lambeth, Summary of Consultants' Final Reports* (HMSO, 1977). Also useful are David H. McKay and Andrew M. Cox, *The Politics of Urban Change* (Croom Helm, 1979), Chapter 7; Paul Lawless, *Britain's Inner Cities: Problems and Policies* (Harper and Row, 1981); Paul Lawless, *The Evolution of Spatial Policy: A Case-Study of Inner-Urban Policy in Great Britain, 1968–1981* (Pion, 1986) and Paul Lawless and Frank Brown, *Urban Growth and Change in Britain: An Introduction* (Harper & Row, 1986). For the problems of London, see Nick Buck, Ian Gordon, Peter Hall, Michael Harloe and Mark Kleinman, *Working Capital: Life and Labour in Contemporary London* (Routledge, 2002); Peter Hall, *London Voices London Lives* (Policy Press, 2007) and Duncan Bowie, *Politics, Planning and Homes in a World City* (Routledge, 2010). The *State of the English Cities* report (two volumes) directed and edited by Professor Michael Parkinson was published by the Department of Communities and Local Government in 2006.

For discussion of the reforms to planning, the promotion of spatial planning, sustainable communities and the public realm after 2001, see Matthew Carmona and Louie Sieh, *Measuring Quality in Planning* (Routledge, 2004); Nick Gallent and Mark Tewdwr-Jones, *Decent Homes for All* (Routledge, 2007); Matthew Carmona, Claudio de Magalhães and Leo Hammond, *Public Space: The Management Dimension* (Routledge, 2008); and Graham Haughton, Philip Allmendinger, Dave Counsell and Geoff Vigar, *The New Spatial Planning: Territorial Management with Soft Spaces and Fuzzy Boundaries* (Routledge, 2010).

Note

1   Ironically, the Winchester saga ended in 1990 with a decision that was anything but respectful of the environment: the new road would cut straight through a National Heritage area.

7 Planning in Western Europe since 1945

The 27 member states of the European Union (EU) offer some instructive comparisons for the planner – both with each other, and still more so with the experience of Britain as outlined in the preceding chapters. In no EU country, in the early twenty-first century, does agriculture make any significant contribution to the gross domestic product (GDP), while industry accounts for one third at most; almost everywhere, services account for three fifths or more of gross value added. Employment in primary production (agriculture, forestry and fishing) in the great majority of the 27 EU countries is less than 10 per cent and in most less than 5 per cent, though the proportion is still 12 per cent in Latvia and 30 per cent in Romania; industrial employment accounts for less than 30 per cent, rising to 39–40 per cent only in Slovakia and the Czech Republic; services thus generally account for well over half of employment and three fifths of GDP.

But in general – with the possible exception of Belgium – the great move off the land and into the cities took place later than Britain's, it took rather different forms and had rather different spatial impacts. When Britain joined the-then European Economic Community (EEC) in 1973, it had only some 3 per cent of its labour force in agriculture, and some 80 per cent of its population was urban – a percentage almost unchanged since the beginning of the twentieth century. But the ten EEC countries at that time still had about 14 per cent of their workers in agriculture, and in general a higher proportion of their people lived in villages and small towns. One key reason for this is that with rare exceptions – the Ruhr coalfield in Germany and the nearby coalfield of southern Belgium, or the similar coalfield area around Katowice in southern Poland – mainland Europe avoided the rapid industrialization which produced the sordid industrial landscapes that once disfigured the Midlands and northern England; coming much later, after the advent of railways and even of electric power, the industrial revolution in these countries affected the existing older cities, so that its effect both on social patterns and on the landscape was less profound.

But these differences should not be exaggerated. The major economic and social trends are as unmistakable in all the Continental European countries as in Britain. Despite strongly protectionist agricultural policies which result from the historic strength of the farm vote (and are still embodied in the Common Agricultural Policy), the figures show just how many millions of workers – especially younger workers – have left the family farms since 1945; the system of peasant farming, which was typical in most of these countries, would not guarantee them the standard of living they expected, so they moved to the cities, eventually leaving ancestral farms in ruins or restored as second homes for townspeople. Especially in the 1950s and 1960s, there were big long-distance migrations of farm workers from the poorer parts of the countryside, especially from southern Italy, to the major industrial areas of Europe. In these reception areas, cities have grown to form the equivalents of the great British conurbations – the *agglomérations* of France, the *Ballungsräume* of Germany. And, after joining the EU in 2004, countries like Poland and Lithuania saw a similar outflow from their poorer still-agricultural regions both to their own capital cities and to the cities of Western Europe – above all to London and other UK cities. A stark contrast has appeared, in all these countries, between the backwardness and stagnation of the remoter rural areas and the dynamic growth – too often accompanied by familiar problems of congestion, high land prices, poor living conditions and pollution – of the major urban agglomerations. But also, since 1980, another problem and another contrast has emerged, long familiar in Britain but newer on the European mainland: though population growth in the EU sharply slowed, producing the threat of population decline in countries like Germany and Spain, and some large and older-industrialized urban areas were losing people, employment in manufacturing contracted so rapidly – by more than 8 million, or 18 per cent, between 1971 and 1981 alone – that by 1997 unemployment in the EU-15 totalled 10.7 per cent, against 2.1 per cent average for the EEC-6 of the 1960s and 3.8 per cent for the EEC-10 of the 1970s; it then steadily fell during subsequent decade of growth, to only 8.6 per cent in 2005, before rising again in the recession at the end of that decade; employment in the enlarged EU-25 was significantly higher, 8.8 per cent. The resulting new contrast was between the nineteenth-century agglomerations, based on coal, iron and steel, heavy engineering and on port activities, such as northern England, the Ruhr, Lorraine, the Belgian–French coalfield, Saxony and Silesia; and others, often ironically much older in origin, which have made the successful transition into twentieth-century high-technology manufacturing and higher-order service functions, such as southern Germany, the Mediterranean coast of France, the Emilia–Romagna region of Italy or south-east England.

These contrasts emerge clearly in the maps of statistical indicators within the community. In the early twenty-first century GDP per head of population ranges from 25 per cent of the EU-27 average (5,800 PPS) per inhabitant in the north east (Romania) to 336 per cent (79,400 PPS) in the UK capital region of inner London – a factor of nearly 14:1. Generally, it is highest in southern Ireland, southern England, the Benelux countries, the Île-de-France and Madrid regions, south-west Germany, northern Italy and the urban parts of Scandinavia. It is conspicuously lower across the recent accession countries of Eastern Europe (Figure 7.1a). With only four exceptions – the regions around the capital cities of Prague (Czech Republic), Bratislava (Slovakia) and Warsaw (Poland) and Malta – all other regions of the new member states and Croatia have a GDP per inhabitant of less than 75 per cent of the EU-27 average. But these regions have been rapidly catching up, both before and after their accession, while large areas of Western Europe have been relatively stagnating (Figure 7.1b). Despite this, unemployment remains stubbornly high in much of the east, but also in other peripheral regions such as southern Italy, southern Spain and Greece (Figure 7.1c). It is small wonder then that there has been migration out of Eastern Europe into the more prosperous West – and this, combined with low natural growth, produces a pattern of very slow overall population growth across Germany and southern Poland (Figure 7.1d).

There is thus a new geography of Europe – a geography of stagnation and growth, of 'have' regions and 'have-not' regions. This geography appears to ignore international boundaries. The major urban agglomerations tend noticeably to have a central location within the European Community; most of them are found within a five-sided figure, the 'Pentagon' identified in the 1998 European Spatial Development Perspective, bounded by Paris, Milan, Munich, Hamburg and London (Figure 7.2a). Interestingly, this figure (the name of which echoes the French self-description of their country as L'Hexagone) has supplanted another in planners' discourse, identified by French researchers in 1989: the 'Blue Banana', a linear, highly urbanized zone including London, Brussels, Amsterdam, Cologne, Frankfurt, Munich and Milan (Figure 7.2b); it differs only by the

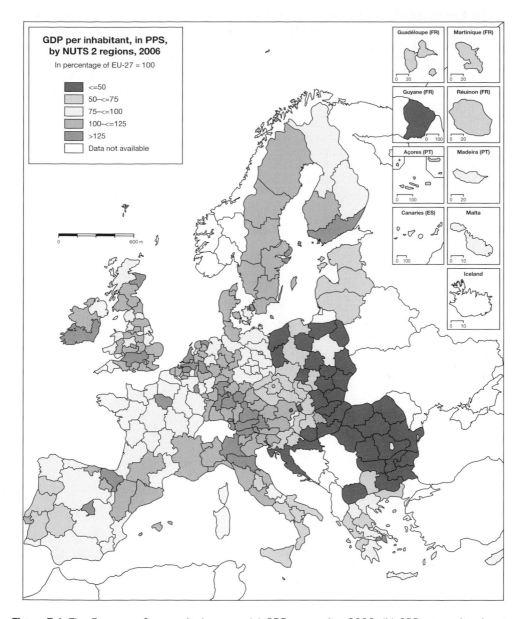

Figure 7.1 The European Community in maps: (a) GDP per capita, 2006; (b) GDP per capita change, 2001–6; (c) population change, 2003–7; (d) unemployment, 2007. The accession of 12 new eastern and southern European nations to the EU in 2004 and 2007 brought widening spatial disparities in economic performance and prosperity. Europe's most prosperous regions are found in a corridor from southern Ireland, through southern England, through the Netherlands and the Rhine Valley of western Germany to central Switzerland and across the Alps to northern Italy, with outliers in Scandinavia and northern Spain. The good news was, however, that these differences were shrinking as regions all around the European periphery – from Ireland to northern Sweden, eastern Europe and western Spain – rapidly caught up. Demographically the picture was different: while Europe's western and southern peripheries showed strong population growth fuelled by in-migration, wide areas of central and eastern Europe, and parts of Scandinavia, were experiencing sharp decline through low rates of natural increase coupled with out-migration. Unsurprisingly, these areas of out-movement were also those experiencing high rates of unemployment.

inclusion of Paris. Conversely, regarding the Pentagon as the new European heartland, the major problem areas are all well outside it on the EU periphery: they include much of midland and northern England, much of Scotland; all of Ireland, north and south; much of southern and western France, below the diagonal line from Le Havre on the Normandy coast down to Marseilles; all of Italy outsider the northern belt; the so-called *Neue Länder* of eastern Germany (the former Communist DDR); all of Spain, Portugal and Greece; and all of the new accession countries of Eastern Europe.

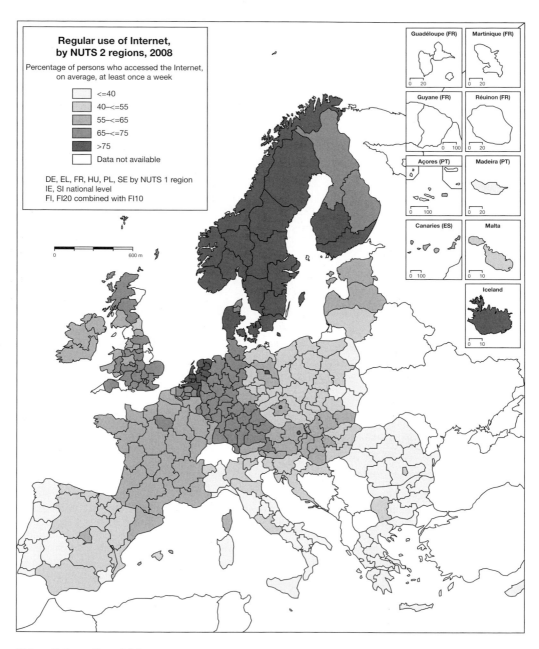

Figure 7.1 continued (b)

There are very good economic reasons for this situation. The *raison d'être* of the agglomerations – Greater London and its surrounding towns, the Randstad of Holland, the Rhine–Ruhr and Rhine–Main areas of Germany, the Paris region – is specialized high-technology manufacture and, increasingly dominant, tertiary (service) industry (Figure 7.3a, b). Both these industries seek locations with large markets and large, skilled labour forces; tertiary industry, and its modern outgrowth, decision-making quaternary industry, demand specialized transportation and marketing services and

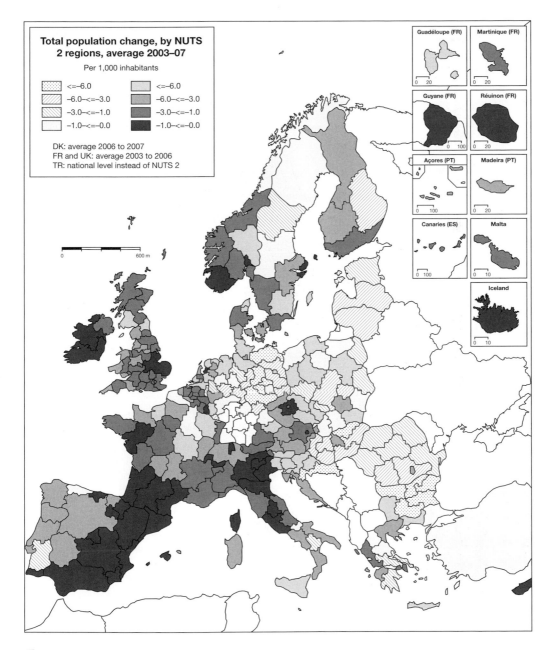

Figure 7.1 continued (c)

increasingly form a complex in order to exploit economies of agglomeration and scale. Goods and also non-material intelligence are increasingly exchanged between these areas. Because powerful forces of inertia work in the location of such activities, they tend to grow where they have been traditionally located: in the old-established trading and governmental centres, which in turn are related to historic trade routes. In the twentieth century as in the Middle Ages, these are heavily concentrated in a relatively small zone of northern France, the Netherlands, western Germany and south-east England, spreading

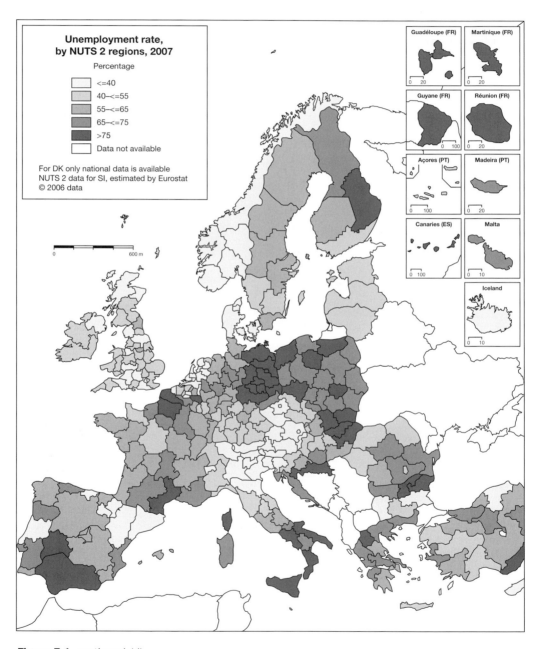

Figure 7.1 continued (d)

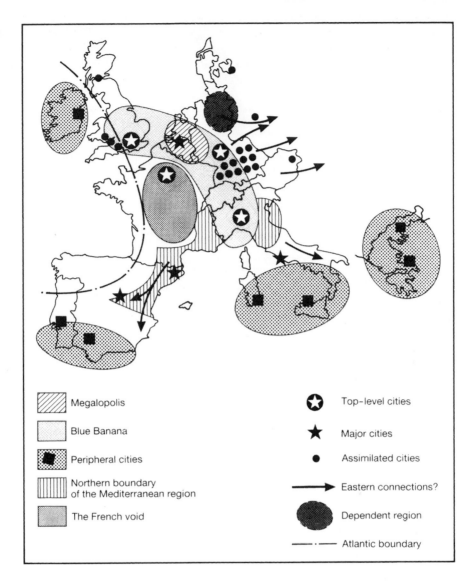

Figure 7.2 (a) The 'Blue Banana' and (b) the 'Pentagon'. These two symbolic, but highly potent, geographical images embody sharply visions of Europe's economic and demographic dynamism over a decade. French researchers, who conceptualized the Blue Banana in 1989, saw a Europe dominated by strong growth along an axis from southern England, up the Rhine Valley to Switzerland and northern Italy, conspicuously bypassing Paris. The EU's 1999 European Spatial Development Perspective embodied a different vision, dominated by a significantly larger central Pentagon whose corners were London, Paris, Milan, Munich and Hamburg. Once again, though, the policy implication was that there were big spatial disparities requiring to be reduced – and the indices (Figures 7.1 (a)–(d) showed this to be substantially true).

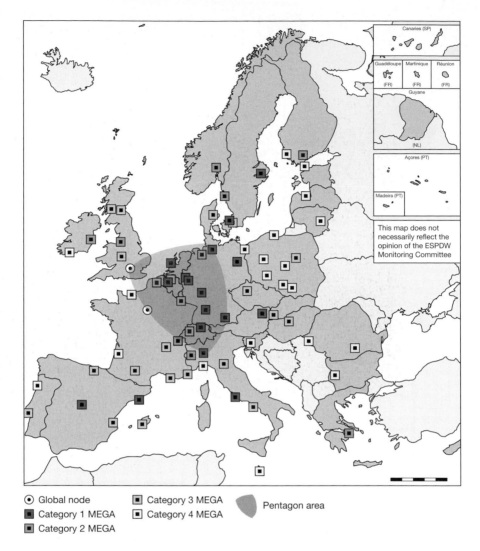

Figure 7.2 continued (b)

in a line southwards up the Rhine and across the Alps to northern Italy. The stagnant rural areas, in contrast, are without exception well away from these major lines of force in European geography. The existence of the EU, with the associated growth in trade in industrial goods and services among its member states, can only reinforce the trend, unless corrective action is taken. Within the areas of urban concentration, however, sharp contrasts have recently emerged between the coalfield and port cities that concentrated too heavily on a few functions which fell into decline, and the more diversified trading cities which have continued to do well. Areas like the northern French–southern Belgian coalfield, the Ruhr area, Lorraine, northern Spain and Lombardy have been afflicted by the decline of their basic industries in the same way as the Midlands and north of England, while the Rhône valley and the Mediterranean coast of France, southern Germany, Madrid and Emilia-Romagna have flourished.

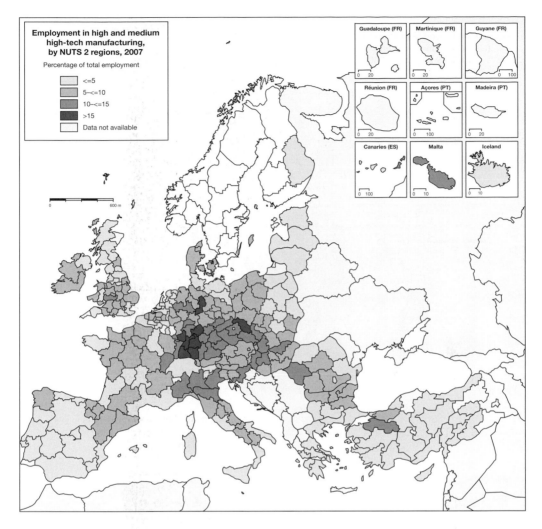

Figure 7.3 European regions: (a) high-technology manufacturing, 2007; (b) business services, 2008. The so-called twenty-first-century knowledge economy is unevenly distributed in Europe. High-technology manufacturing is heavily concentrated in the heart of the EU, in southern England, the Benelux countries, southern Germany and northern Italy. Business services are much more widely distributed, but with particular concentrations in southern England, the Benelux countries, western Germany, western and southern France, and central Italy.

The result is a continuing, but also an increasingly complex, pattern of regional differentiation. Broadly, it continues to be true in the 2000s, as in the 1960s and 1970s, that in large tracts of France and Germany and Italy (and indeed Spain, Portugal, Sweden and Finland), as well as across most of Eastern Europe, that the population is too thin and scattered to support modern services, and the many market towns are working at much less than the scale for which they were intended; and that from the urban agglomerations come similar stories of housing shortages, traffic congestion and long journeys to work; rising land prices and land shortages; public services that cannot cope. But some rural areas have become reception areas for people and activities decentralizing

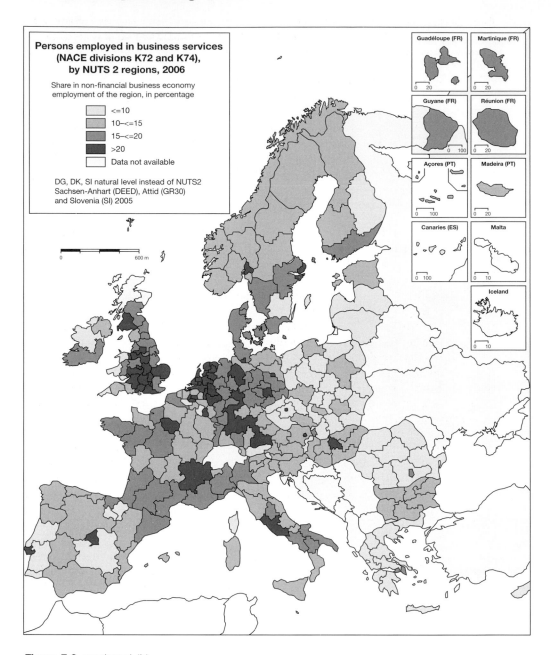

Figure 7.3 continued (b)

from the agglomerations, thereby acquiring a new lease of life; and, as already seen, some of the specialized industrial agglomerations have suffered severe problems of economic adaptation. So the regional map of Europe looks more complex than it did, and the appropriate measures have changed too.

The problems cut across national frontiers; and, as the European Union increasingly practises European-wide assistance and controls the actions of its governments, so do many of the resultant solutions – whether in south-east England or the Île-de-France, the Rhine–Ruhr area of Germany or north-east England, Brittany or Ireland, Andalusia in Spain or the Italian Mezzogiorno, Saxony or southern Poland. But there remain important differences between one country and another. Until the French administrative reform of 1982, the strongly centralist tradition of French public administration contrasted with the federal system of Germany. Controls over land use have been more effective in north European countries, such as Scandinavia, Germany and the Netherlands, than in Italy.

To grapple with these problems, each country has developed its own individual set of regional policies. But from the start of the original six-member EEC in 1957, member states have been compelled to adjust these policies to overall Community requirements which forbid artificial impediments to competition. And progressively, the Community (which became the EU with signature of the Maastricht Treaty in 1992 and its ratification in 1993) developed its own overall measures to aid regional development. The oldest, the European Coal and Steel Community, was set up as long ago as 1951 – six years before the Treaty of Rome which brought the original six-member EEC into being – to modernize and rationalize production in these old basic industries, and to help retrain unemployed workers. The Community worked well until the crisis of the late 1960s, which progressively overwhelmed it. The Social Fund, set up under the Rome Treaty, sought to improve employment opportunities by such devices as retraining, mobility allowances and aid to regions in economic transition – an implicit regional policy, though such policy formed no part of the treaty. Similarly, the European Investment Bank was set up in 1958 in part to help less developed regions; during its first 15 years 60 per cent of its funds went to Italy, especially the Mezzogiorno, but later – especially after the entry of the UK in 1973 – it has turned to declining industrial areas and Britain, France, Ireland and Greece have benefited too. The European Agricultural Guidance and Guarantee Fund contains elements to improve infrastructure and support integrated land development programmes (Clout 1986b; 40). The European Regional Development Fund (ERDF), created only in 1975, is most directly important with a total budget of €277 billion between 2007 and 2013. The ERDF, the European Social Fund and the Cohesion Fund are brought together into a regional development grant package, totalling €347.4 million: 44 per cent of EU funding, almost precisely equal to the agriculture budget (43 per cent). Of this total, 81.5 per cent will be spent on the convergence objective – and thus will be heavily concentrated on the poor regions in the new member states of east central Europe. Since the highest level of assistance is granted automatically to regions whose per capita GDP is 75 per cent or less of the overall EU average (formerly known as 'Objective 1', now as 'Convergence Objective'), virtually all this aid was shifted from the poorer-performing regions of Western Europe to the new Eastern European members (Figure 7.4a and b).

Within the different EU countries, Hugh Clout's verdict is relevant: most systems of regional aid have been geared to job-creation schemes and have historically emphasized manufacturing rather than the services sector of the economy. Most embody a mixture of direct incentives (financial inducements), indirect investments in infrastructure and housing, direct disincentives in the form of regulations or licensing of new jobs, and

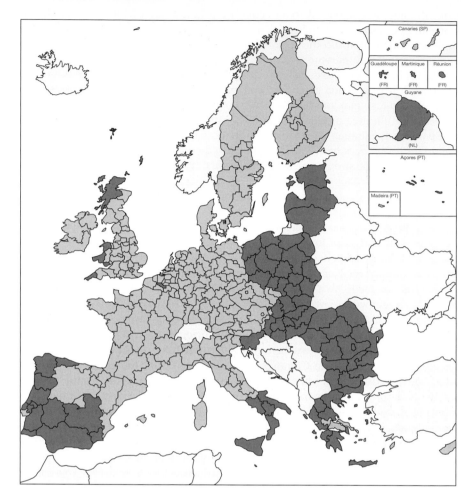

Figure 7.4 EU Convergence and Competiveness areas, 2007–13. In the new map of EU regional aid (a), a conspicuous feature is the effective disappearance of the old Objective Two regions, chiefly old industrial areas receiving assistance to make the transition to the new service-based knowledge economy. The old Objective One regions (b), still defined in terms of low per capita GDP, continue to receive generous aid but are now dominated by the formerly Communist new accession states of Eastern Europe.

indirect disincentives in the form of taxes and similar devices. Because each of the European countries shows at least some unique features, their problems are best discussed separately, with a general summing up at the end.

French planning

There are many reasons for starting with France. The country shows an extreme version of the centre–periphery contrast, resulting in an acute problem of planning at what we have called, throughout this boom, the national/regional scale. But the very size of the Paris agglomeration throws up additional questions of planning at the scale of the city

Figure 7.4 continued (b)

region: the regional/local scale. At both scales, the French have shown remarkable inventiveness in developing new organizations and new techniques of planning. Indeed, at the national/regional scale they have developed a planning apparatus which is unparalleled, in its comprehensiveness and its sophistication, in the developed world.

The geographical and historic background to the problem is a highly individual one. In the nineteenth century, France never experienced the rapid population growth typical of other advanced countries. Large-scale industry, with the exception of a concentration in the north, around Lille near the Belgian frontier, failed to develop on any scale. Instead, because of the strong tradition of centralization in French life, Paris grew apace while other parts of the country stagnated and even declined. Paris came to dominate the economic and social life of the country to an unusual degree.

Since the Second World War the demographic situation has been revolutionized; population has rapidly grown, but in the process it has concentrated further in the urban areas and above all in Paris. Two thirds of the population was urban in the 1960s, as opposed to only one quarter a century before. A significant part of this urban population, and of the total population growth, was concentrated in the northern and eastern parts of France; south and west of the critical line from Le Havre to Marseilles, there was a contrasted rural landscape of stagnation and decay. By the early 1960s the Île-de-France, occupying 2 per cent of the area of France, had 19 per cent of its population and 29 per cent of its industrial jobs; an even smaller area within it, the city of Paris proper, had one quarter of the nation's civil servants, one third of the higher education students and nearly two thirds of all the commercial headquarters. In contrast, the rural west, with 55 per cent of the area, had 37 per cent of the population and only 24 per cent of the industrial jobs.

A remarkable book published in 1947, right at the beginning of the postwar reconstruction period, first drew attention to the problem. *Paris et le désert français*, published by a young geographer, Jean-François Gravier, argued that the contrast was rooted in an accident of history and not in economics; technological innovations, such as widespread electric power and motor vehicles, Gravier argued, could promote dispersed industrial development in the countryside and reverse the trend.

Gravier's book had immense influence and soon brought practical results, for in 1946 France had embarked on an ambitious experiment; under the direction of Jean Monnet, the country tried to develop a system of economic planning based not on state ownership of all resources (as in Soviet Russia during the 1920s), but on a mixed economy where about half the total investment was in private hands. In the early years of the plan, there was little interest in questions of the geographical distribution of investment or of economic activity generally. But from the mid-1950s, as economists took an interest in these questions, the regional element became an increasingly important part of the plan. Already in the early 1950s special state funds were created for regional development, though these were outside the plan process; from 1955 the central plan agency (the *Commissariat général au plan*) was given regional responsibilities. In the same year, a decree established that government approval would be necessary for new factory building or reconstruction in the Paris region and thereafter the capital's proportion of new industrial building did fall. Appropriately, at the smaller scale of the city region, a 1960 plan for Paris (the so-called PADOG) proposed a stop on the future physical growth of the agglomeration.

By this time the process of integrating regional and national planning was becoming increasingly sophisticated. The resulting structure, as fully evolved in the mid-1960s, was a highly complex one (Figure 7.5). Essentially, the *Commissariat général au plan* (CGP) worked through a regional arm, the *Délégation a l'aménagement du territoire et l'action régionale* (DATAR), established in 1963, which coordinated regional agencies and administered regional development funds, and which was responsible directly to the prime minister. The country had already been divided, in 1955, into 21 economic planning regions (Figure 7.6), to which Corsica was added in 1970, consisting of groups of the *départements*, the basic administrative units of France; each region had a *préfet*, an official of the central government (eliminated in 1981), assisted by a regional conference of officials and a regional commission of appointed experts from areas like industry, trade unions and universities. Together the regional and the central planning machine prepared regional sections (*tranches opératoires*) of the plan, through an elaborate process of refinement conducted between centre and region. In fact, as Figure 7.6 demonstrates, the relationship and responsibilities – especially at the centre – were far from clear, they resulted only partially from the rational thought for which the French are renowned, and rather more from interdepartmental rivalries and suspicions. That is why there are so many parallel bodies, and so few vertical responsibilities in the chart.

The aims of this elaborate machinery were at bottom mundane enough: to redistribute employment in manufacturing and latterly in tertiary activities; to promote the

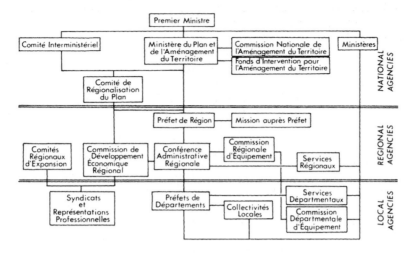

Figure 7.5 The administrative structure of French regional planning. To integrate national and regional planning, the French had to develop a complex structure during the 1960s. The system of *préfets* was abolished by the Mitterrand government in 1981.

modernization of farming and the creation of non-agricultural jobs in rural areas; and to enhance commercial functions, modern branches of manufacturing and higher education in the *métropoles d'équilibre*, designated in 1963 (Figure 7.6). Within the Paris region projects for industrial expansion are scrutinized.

Yet another system exists to promote and coordinate plans at city-region level. In 1966 metropolitan plan organizations (each known as an OREAM) were set up for the six major urban regions of Lille–Dunkerque, Rouen-le Havre, Nantes–Saint Nazaire, Lyon–Saint Etienne, Marseille–Aix and Nancy–Metz; while a central planning group in Paris, set up two years earlier (known as GCPU, for *Groupe central de planification urbaine*), advises ministers on planning questions at this level. The organization for the Paris region is of course rather special; set up in 1961, it has provided a model for the other regions. Here, a full-time regional *préfet*, assisted by a team of civil servants, chaired a board consisting of local government representatives from the region. There is also a research association, corresponding to the OREAM in the other major urban areas; called the *Institut d'aménagement et d'urbanisme Île-de-France* (IAURIF), it has acquired a considerable international reputation for the quality of its studies. At the same time, in order to bring the government of the Paris region into line with the realities of its great size and complexity, the number of *départements* within the region was increased from three – out of an old total of 90 for the whole country – to eight, thus raising the national total to 95. In the reforms introduced in 1982 by President Mitterrand, the Île-de-France region, like all the other French regions, was given enhanced autonomy including independent funding capacity, which it has used vigorously to promote ambitious investments in public transport.

Elaborate machinery – sometimes confusingly elaborate – is thus one of the outstanding features of French regional planning at both national/regional and regional/local scales. The critical question must be what this machinery has achieved. At the major scale of the relationship of the regions to the national economy the policy of trying to restrict the growth of Paris has been retained; but it has been modified. In contrast to the 1960 plan which tried to put an absolute stop on the physical growth of Paris, a later 1965 plan

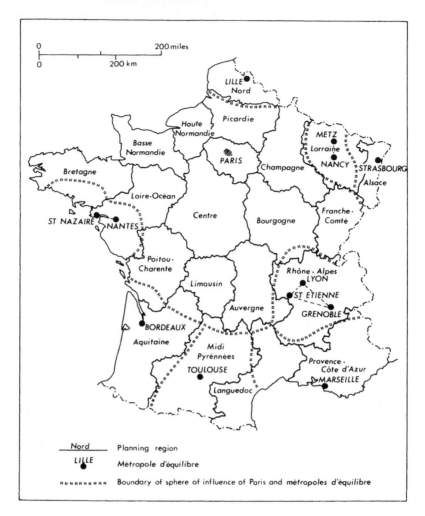

Figure 7.6 French planning regions and the *métropoles d'équilibre*. French regional planning is carried out through 21 planning regions which are aggregations of departments. To try to achieve more balanced growth and avoid over-concentration on Paris, the objective is to concentrate investment in 'balancing metropolises' based on the major provincial cities.

(the so-called *Schéma directeur*) assumed a continuing high rate of growth of population (4 per cent per annum, giving more than a doubling of population in 20 years), but slowed down the planned rate somewhat in the interests of the other major urban regions. The objective has been to slow down the rate of migration to the Paris region, first by careful localization of government investment, and then by guiding private investment through the provision of public infrastructure. Especially important here was the designation, in 1963, of eight *métropoles d'équilibre* ('balancing metropolitan areas') designed deliberately as counterweights to the capital (Figure 7.6). Carefully selected on the basis of the major provincial centres of population, they are designed to act as centres of economic development for their respective regions. The first, based on the northern cities of Lille, Roubaix and Tourcoing, is designed to help the regeneration of an old industrial area based on coal and textiles. The second, Nancy–Metz in Lorraine, and the third, Strasbourg in Alsace, are based on quite prosperous eastern industrial areas. The fourth, Lyon–Saint Etienne, includes a problematic coalfield area and serves the poor marginal hill-farming area of the Massif central. The fifth, Marseille–Aix, includes both the rapidly developing industrial area of the Lower Rhône and the southern slopes of the problematic Massif. The sixth, Toulouse, is a very important centre of industrial development in the south west, best known for its aircraft building complex which built the first Concorde. Not very far away on the west coast is the seventh centre, Bordeaux; the eighth, Nantes–Saint-Nazaire, is farther north on the same west coast and is intended to as the springboard for the development of Brittany.

Since the late 1960s successive French governments have systematically sought to divert public investment into these poles, thus strengthening their economic potential and acting in turn as a device to attract private capital. To take two examples: the country's higher educational system, previously dominated by the Sorbonne and the *grandes écoles* (professional schools) in Paris, has been profoundly modified by the expansion or establishment of both kinds of institution in the *métropoles* – as also by the decentralized expansion of the historic University of Paris into 13 separate campuses, the majority in the suburbs. The development of the motorway system in the 1960s and 1970s, and that of the *TGV (Train à Grande Vitesse)* system in the 1980s and 1990s, has clearly been dominated by connections between Paris and the *métropoles*: thus the *TGV Sud-Est* of 1981 linked Paris, Lyon–St Etienne and Marseille–Aix; the *Atlantique* in 1989–90 connected Paris, Nantes–Saint Lazaire, Bordeaux and Toulouse; the *Nord* in 1993 connected Lille–Roubaix–Tourcoing; in 2007, the *Est* hooked in Nancy–Metz and by 2016 will be extended to Strasbourg (Figure 7.7 and Plates 7.1, 7.2a and 7.2b).

This is a dramatic and bold policy, which corresponds fairly well to the realities of French geography; outside the Paris region, France is less urbanized than Britain or Germany and the urban population is heavily concentrated into the eight regional centres, so that these are the logical places from which to generate regional economic development. But there are two snags. One is that all the regions contain large (and often thinly populated) rural areas which are well outside the sphere of influence of these centres; to help them it would also be necessary to develop other 'poles of growth' based on smaller centres, but that runs the risk of spreading investment too thinly. The other problem is that investment in the *métropoles* also has to compete with investment in Paris. Though to many provincial French people Paris seems to have a disproportionate share of everything, the fact is that for many decades the capital city's infrastructure was running down through under-investment. After the great burst of investment under Baron Haussmann in the 1850s and 1860s, there was relatively little new house building; after the construction of the new boulevards by Haussmann at that time and the building of the *Métro* in the early years of the twentieth century, the transport system also suffered the effects of low investment. To make the city more efficient and more liveable required a massive dose of investment.

This is underlined when one looks more closely at the problems of planning at the regional/local scale within Paris itself. To try to make up for the backlog of investment, the 1965 plan suggested the creation of eight new cities, strung out along two parallel axes on either side of the Seine, east and west of Paris: the first, 55 miles (88 kilometres) long, south of the river from Mélun to Mantes, the second, 45 miles (72 kilometres) long, north of it from Meaux to Pontoise (Figure 7.8a). These would nearly double the size of the existing built-up area within a 35-year period up to the end of the century. To service all this would demand 540 miles (869 kilometres) of new highways and 156 miles (251 kilometres) of an entirely new regional express rail (RER) system (the first parts of which were opened in 1971, and which was effectively completed by the end

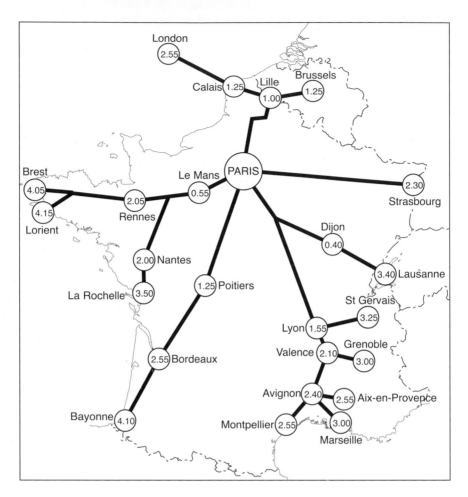

Figure 7.7 The TGV system; times from Paris. The trains have massively cut times and shrunk distances in the urban core of Europe, becoming the natural mode of travel for journeys up to about 500 miles (800 kilometres).

of the 1990s, with four lines complete and a fifth partially open). Also involved was the expensive renovation of existing centres within the urban fabric of Paris, such as those at La Défense and Nanterre (the first to be begun and largely completed by the early 1970s, but then massively extended through work which continues), Saint-Denis, Bobigny, Créteil, Versailles and Choissy-le-Roi/Rungis (the site of the new markets of Les Halles, close to Orly airport). All this was necessary, of course, not only to make up for the deficiencies of the past, but to cater for a population growth that might take the population of the region from 9 to 14 million within 35 years.

Both scales of planning had a common theme – the attempt to break the concentration of economic life at the centre, by developing a full range of economic opportunities, and of social and cultural facilities, in a number of urban counter-magnets. From this viewpoint the new cities of the Paris region, some of which were originally planned with target populations of up to 1 million inhabitants, would perform essentially the same role as the *métropoles d'équilibre* in the provinces.

Plate 7.1 TGV and tram at Grenoble, a dynamic high-tech city in the French Alps. One of the fastest trains in the world, the *Train à Grande Vitesse* (high-speed train) runs at speeds up to 205 miles per hour (325 kilometres per hour) between Paris and other European destinations along the European high-speed rail network. Here, as at Grenoble and other French cities, it is increasingly integrated seamlessly with the city's new tramway system, providing integrated and convenient public transport that effectively challenges the private car.

In the event, then, policies had to be adapted to changing circumstances – above all, demographic ones. A falling birth rate meant that the Parisian regional population target was cut from 14 to 10–11 million, the latter figure representing only a marginal increase on the total already achieved by 1978. The number of new towns was cut to five: Cergy-Pontoise and Marne-la Vallée on the northern axis, St Quentin-en-Yvelines, Evry and Melun-Senart on the southern. Progress on them – in housing, industry and above all offices – was at first slow, but accelerated spectacularly from the late 1970s. During the five years 1977–82 they actually achieved more than 90 per cent of the total population growth in the *Région Île-de-France*; by the end of the 1990s, they had a combined population of 743,000, one in 15 of the entire population of the region. Progress on the RER rapid transit links continued as planned, and the entire original four-line system was completed in the 1990s, with partial opening of a fifth line; but construction of the equally ambitious circumferential motorways was delayed by planning and environmental problems, culminating in a decision to route the most difficult link – the western sector of the A86, the middle ring – in two long deep-level tunnels, built by the private sector as a toll road, due to be completed in 2010. Though shopping decentralized rapidly to the suburbs, and deficiencies in public services there were made good at remarkable

(a)

(b)

Plate 7.2 (a) Reconstruction of La Défense, Paris. Located just outside the limits of the historic city of Paris, La Défense was one of the biggest pieces of reconstruction in Europe in the 1960s and 1970s. It contains offices, homes, a station on the *Métro* and a highway interchange. (b) The western extension of La Défense, a continuation of the great linear axis of Paris begun by Louis XIV, with public space over an underground motorway.

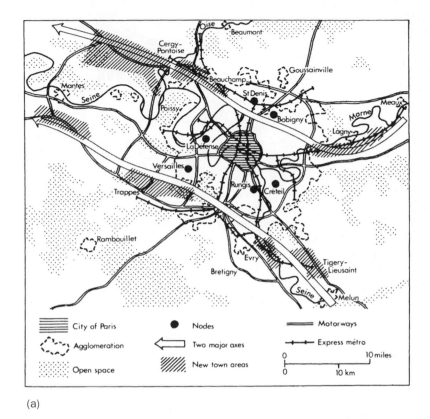

(a)

Figure 7.8 Paris regional plans. (a) The *Schéma Directeur* of 1965; to allow for the projected growth of the metropolis from 9 to 14 million by the year 2000, the plan provided for a number of major new cities grouped along two major axes of development, one north and one south of the River Seine. New motorways and express rail links were to serve the new developments. The plan was scaled down, but was implemented. (b) The *Livre Blanc* of 1990; the stress here was less on population growth than on creating strong poles for European-scale service industries. (c) SDRIF 2008; the latest iteration of the Spatial Development Plan (*Schéma Directeur*) embodies three principles: (1) to develop a more compact, denser city to respond to the challenge of housing demand and to respond to climate-energy constraints; (2) to develop the region's urban ambience and quality of life so as to promote its economic potential and international attraction; (3) to protect regional biodiversity, enhance the quality of its agricultural and natural areas, and guarantee the coherence of the regional open-space system.

speed, the region's fast-growing office sector showed a notable tendency to spread westwards, first into the giant La Défense scheme (Plate 7.3a, b), and then into the adjacent inner suburbs along the Seine where large tracts of land had been released through factory closures. The two western new towns, Cergy-Pontoise and St Quentin-en-Yvelines, benefited; on the other side of Paris, Marne-la-Vallée was boosted by the giant Disneyland Paris theme park at its eastern end, directly connected to a station on the TGV line bypassing Paris, and by a university campus. So, far from becoming a polycentric city, Paris has tended to remain polarized between the centralized business city and the suburban dormitories, and between the affluent west and lower-income east. Nevertheless, the scale of the achievement should not be denied. Paris experienced a major physical restructuring; the paradox is that in the process, its traditional social and economic structure had, if anything, been intensified.

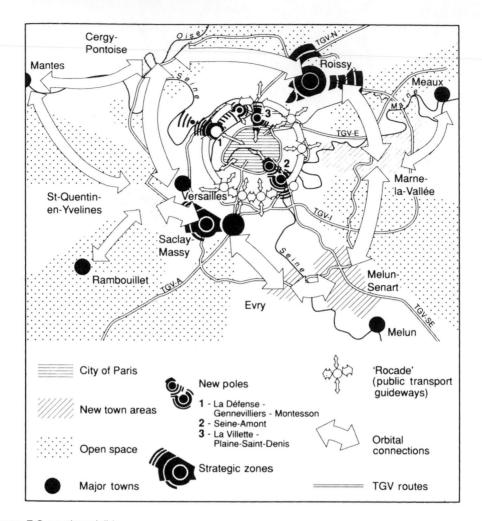

Figure 7.8 continued (b)

In the 1990s, undeterred, the central government and the regional authority – now elected – began work on a new plan for the next quarter-century. This plan, approved in 1999 and revised in 2008, no longer sought to accommodate huge population growth: the region's population, 10.9 million in 1999, was projected to be almost static over the following quarter-century – though in the event, growth in the 1990s was higher than this. Its main new features were first an unabashed concentration on competitive economic development to rival other great world cities, and second a major emphasis on restructuring inner areas, both at the corners of the historic city and in neighbouring zones immediately outside it, to accommodate large-scale deindustrialization and the shift to the service-based economy.

The main spatial emphasis of the new plans is on five giant poles of urban development – three around existing urban locations in the middle ring (La Villette and the Plaine St-Denis in the north, La Défense–Genevilliers–Montesson in the west and the Upper Seine valley in the south east) and two at the edge of the agglomeration (Roissy–Charles

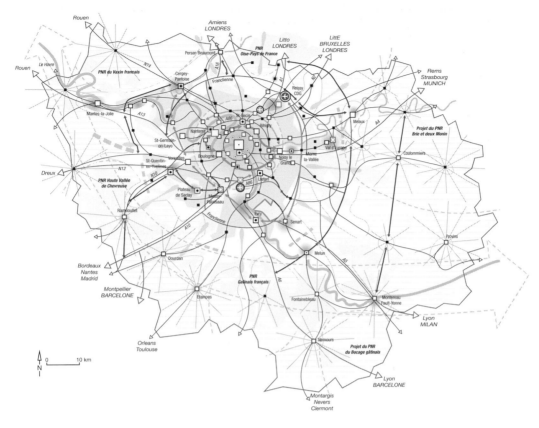

Figure 7.8 continued (c)

de Gaulle airport in the north, Saclay–Massy in the south west). These would accommodate the growing service sector activities – financial and business services, universities, hospitals and cultural centres. These and the five new towns would be linked by new motorways and public transport links, including a new public transport 'rocade' linking the three inner poles; a vast green area would be preserved, both within the agglomeration and outside it (Figure 7.8c).

By the early twenty-first century the Île-de-France region had grown to house 11.4 million people: nearly 19 per cent of the French population, living on only 2 per cent of its area. And it was growing strongly, by 440,000 people between 1999 and 2005, so that projections suggested a population of 12–13 million by the year 2030. The associated growth of households, 50,000 a year, far exceeded the expansion of the housing stock, with the predictable result that housing costs were rising more than twice the rise in average incomes, with particularly dire effects for lower-income residents; there was growing polarization and income disparity between rich and poor, similar to the phenomenon in other great global cities. Employment too was rising and was increasingly dominated by the service sector, which employed no less than 83 per cent of the entire workforce; just under half, a smaller proportion than in 1999, were working in the city of Paris and the adjacent Hauts-de-Seine *département* to the west, which now constituted the new central business district of Paris. Here, the La Défense complex had

(a)

Plate 7.3 (a) Euralille: view of the World Trade Centre next to the Lille Europe Eurostar/TGV station. (b) Lower-level view of Euralille with shopping centre, offices and hotel. Planned on abandoned military land around the new high-speed line from London via the Channel Tunnel, Euralille is a spectacular example of the French approach to planning integrated *grands projets* around new transport links, here in order to trigger urban regeneration and regional development in a deindustrialized former coalfield area of northern France.

continued relentlessly to grow, effectively becoming the major business complex of Paris, supplemented by huge public works: the undergrounding of the A14 motorway and the enlargement of the public transport interchange. The restructuring of the depressed former industrial zone in the north (la Plaine de St-Denis) had proceeded more slowly, despite a major boost through the location here of the new national stadium and the construction of one of the first stages of a new orbital public transport link (*L'Orbitale*) through the inner suburbs; and similarly with the south-east corner of the city (Seine–Amont) despite a similar boost in the form of a new national library which became mired in construction and design problems.

Perhaps unsurprisingly, the 2008 revision of the *Schéma Directeur* essentially developed the same principles as in 1999. The central business core was to expand, particularly into adjacent areas which had been the principal seats of Parisian manufacturing until its effective disappearance in the 1980s and 1990s: the north-east corner of the city and the much larger adjacent area in the Seine-St-Denis *département* to the north, the adjacent zone to the west, at the northern end of Hauts-de-Seine, and stretching northwards towards the Le Bourget airport; and the south-east corner of the city, stretching southwards towards Orly airport. And there was strong emphasis on providing new housing in

Plate 7.3 continued (b)

brownfield developments well served by public transport, especially in the old industrial zones. The plan called for 60,000 new dwellings/year, of which two-thirds would go into already-developed (brownfield) zones: 30,500 in the central core of the agglomeration. Here, there would be a marginal increase in average density (by 9 per cent, from 80 to 87 dwellings per hectare), with much lower densities (34 dwellings per hectare) proposed for greenfield development (Figure 7.8c).

To open up these areas, the 2008 plan further developed its predecessor's central concept of strong orbital public transport links: an orbital light rail system, running either at the southern edge of the city (the Boulevards des Maréchaux) or just outside it (La Défense–Boulogne–Issy) was to be supplemented by the Arc Express, a mainly underground rail link, running through the outer zones of the agglomeration, would connect suburban locations with development potential, particularly new poles for tertiary economic activities. In addition, there were ambitious plans for new outer orbital TGV links south and west of the agglomeration, with a new TGV station at La Défense. All this was to be balanced by inserting more green and blue nature into the core, plus a green belt, protecting peri-urban agriculture.

Thus Paris, successfully adapting from an industrial past to a post-industrial future, will retain its national dominance of the top-level service functions. And, though the contrast can no longer be fairly described as one between Paris and *le désert français*, the deep regional division of the country continues to express itself in the statistical indicators.

The German experience

In the Federal Republic of Germany, as in France, the same contrast is evident. The two formerly separate countries that were reunited on 3 October 1990 were, and to an extent still are, totally different in important ways. The five new eastern *Länder* or states, admitted to the Federal Republic of Germany on that date, had been under a communist regime, the German Democratic Republic, for 40 years. Constituting 30 per cent of the land area of the combined state, they contributed only 21 per cent to the 82 million population total; only one, Saxony, exceeded the average population density for the entire country. With abnormally low birth rates, as soon as the borders were open they began to suffer heavy population losses to the more affluent western German states. At the start of the 1990s their gross domestic product (GDP) per capita was about half the national average; after severe deindustrialization in the years immediately after reunification, as inefficient old factories were unable to compete, at the end of the 1990s unemployment stood at between 14 and 20 per cent, against a national average of just under 9 per cent. By the early years of the twenty-first century differentials had narrowed, and all eastern German regions are now all above 75 per cent of the EU average, but typically GDP per capita in most is only around 70 per cent of the levels in the west, with unemployment over 10 per cent against 8 per cent or less in much of the west.

The addition of large, sparsely peopled areas marginally reduced the degree of urbanization. At the start of the twentieth century some 45 per cent of Germans lived on only 22 per cent of the total land area in so-called central areas, highly urbanized and highly connected by excellent roads and public transport. These include the major urban agglomerations of Hamburg, Bremen, Rhine–Ruhr, Rhine–Main, Stuttgart and Munich. One quarter lived in so-called 'intermediate areas': smaller cities but still well connected. Another quarter lived in thinly populated rural areas, which occupied no less than 58 per cent of the land area. While in the old Federal Republic the major urban areas are highly concentrated in two axes – one from Hamburg and Bremen via Hannover to the Ruhr, the other, forming part of the 'Blue Banana', up the Rhine from the Ruhr via Cologne and Frankfurt to Mannheim and then across to Stuttgart and Munich – the pattern in the east is different: there are two distinct agglomerations, one, with 3.4 million people, in the reunited Berlin, the other, with about 1.5 million, in the so-called 'Saxon Triangle' of Leipzig–Halle, Chemnitz–Zwickau and Dresden (Figure 7.9). But most notable is that in the west large areas are close to agglomerations or have no severe development problems, in the east large rural areas – some of them near the agglomerations, most more distant – are classed as problematic. In the west these are remote upland areas with bleak climates and rather poor agriculture, like parts of the Eifel near the Belgian border on them west, or the Bohemian and Bavarian Forests (Böhmer Wald and Bayerischer Wald) against the Czech border on the east. But in the east they include very large tracts of land right across the territory of the new *Länder*, from the Baltic coast in the north to the Polish border on the south east.

The contrast between the great agglomerations on the one hand and the more remote rural areas on the other is the outstanding feature of the geography of Germany. But Germany shares with Britain the feature that not all its major urban areas are equally prosperous. The greatest of them all – the Rhine–Ruhr district, with 11.7 million people concentrated into an area of 11,467 square kilometres – suffered badly from the decline in demand for coal from the 1960s: employment in coal fell by more than two thirds between 1956 and 1972, and has continued to fall since then. Conversely, the more consistently prosperous urban areas have been those in the southern half of west Germany, such as Rhine–Main (Frankfurt–Wiesbaden–Mainz), Rhine–Neckar (Mannheim–Ludwigshafen–Heidelberg),

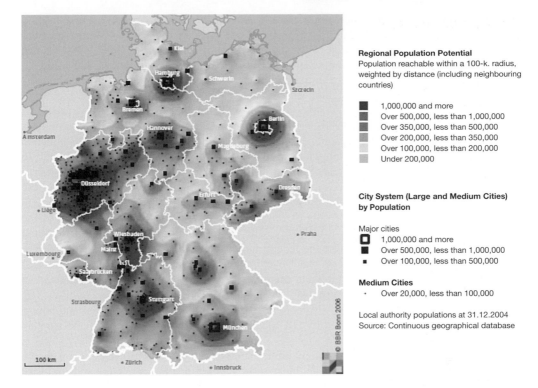

Regional Population Potential
Population reachable within a 100-k. radius,
weighted by distance (including neighbouring
countries)

■ 1,000,000 and more
■ Over 500,000, less than 1,000,000
■ Over 350,000, less than 500,000
▨ Over 200,000, less than 350,000
☐ Over 100,000, less than 200,000
☐ Under 200,000

City System (Large and Medium Cities)
by Population

Major cities
☐ 1,000,000 and more
■ Over 500,000, less than 1,000,000
▪ Over 100,000, less than 500,000

Medium Cities
· Over 20,000, less than 100,000

Local authority populations at 31.12.2004
Source: Continuous geographical database

© BBR Bonn 2006

Figure 7.9 Germany population potential. This map shows the population accessible within a 100-kilometre radius, and is thus a graphic illustration of access to urban labour forces and services. It shows that the German space economy is effectively dominated by the urban fields of only a dozen major metropolitan centres. Large, thinly populated rural areas of north-eastern, central and south-eastern Germany suffer from poor accessibility that impedes their development potential.

Stuttgart and Munich, where faster-growing newer industries, many of them displaced from the east at the end of the war, have tended establish themselves. Just as in Britain we talk of the north–south divide, so in Germany they talk of the *Nord–Süd Gefälle*, the north–south gradient. But this has now been joined by an even more severe west–east gradient. The major cities of eastern Germany were industrially decimated in the years immediately after reunification, typically losing as much of as four-fifths of their manufacturing employment: their factories were too old, too poorly equipped and too poorly managed to compete. Though the major cities have made major efforts to restructure economically, with new service industries – offices, universities, conference and exhibition facilities – the inevitable result is a pattern of sharp contrasts: some parts of the city appear spruce and prosperous, others still careworn and clearly deprived.

Germany's urban areas are much smaller than London or Paris: Berlin with its 3.4 million people is much smaller than either, while the giant Rhine–Ruhr agglomeration is essentially an aggregation of separate medium-sized cities. So they do not have the same problems of congestion and overloaded services; in this sense, Germany has been lucky in its decentralized pattern of urban growth. But in the Ruhr above all there are acute problems of regional/local planning. The decline of basic industry there makes it more difficult to provide the revenue to grapple with questions of traffic congestion, long work-journeys, lack of green space and, above all, air and water pollution. The poisonous state of the river Rhine has become an international problem, since the Netherlands must draw much of its water supply from it. Air pollution from the heavy concentration of chemical and metal plants is a problem not only in the Ruhr – where postwar developments have often been badly sited in relation to residential areas – but also in the low-lying Rhine–Main industrial area between Frankfurt, Wiesbaden and Mainz.

The problems, therefore, are similar to those of France, though the different geography of the countries means that they express themselves differently. A more important difference lies in the administrative tradition, which determines the way the problems are treated. France until the 1980s was a highly centralized country in which the provinces were administered from the capital through a system of civil servants (*préfets*) established in each administrative division. Postwar West Germany has been a federal republic in which basic administrative responsibility for most aspects of home affairs is given to the constituent states, or *Länder*. (They vary greatly in area and population, from the city states of Bremen and Hamburg at one end, to the state of Nordrhein-Westfalen with its 15 million people and the state of Bavaria with one quarter of the total area of the republic at the other. Berlin, a third city state, is an additional special case.) Regional planning is no exception; under the federal law on the subject, passed on 8 April 1965, the Federal Programme of Regional Development, formulated in 1975, lays down that a fundamental aim is to develop and organize the nation so that equal conditions exist everywhere, but the major responsibility to ensure this is given to the *Länder*. The *Gemeinden*, or municipalities, also enjoy a considerable degree of financial autonomy, much greater than that enjoyed by UK local authorities, which gives them freedom in the implementation of policy.

The difficulty is that, as so often, administrative boundaries do not conform to the realities of administrative problems. Most of the *Länder*, it is true, are large and comprise reasonably clearly one or more urban regions. Thus the *Land* of Hessen focuses quite naturally on the Rhine–Main urban area; Baden-Württemberg has a natural focus in the Stuttgart–Heilbronn region; Nordrhein-Westfalen has a heartland in the great Rhine–Ruhr industrial area; and Saxony (Sachsen) is based on the triangle Leipzig–Chemnitz–Dresden. But other *Land* boundaries are by no means as convenient. The most anomalous is around Hamburg, where the tightly drawn boundaries of the city state put the suburbs north of the Elbe in the *Land* of Schleswig-Holstein, and those to the south of the Elbe in the *Land* of Niedersachsen (Lower Saxony). Equally anomalous is the Rhine–Main area, one of Germany's economic heartlands, which is divided among three *Länder*: Hessen (where the largest city is Frankfurt but the capital is nearby Wiesbaden), Rhineland-Pfalz (where the capital is Mainz) and a small portion of Bavaria, where the capital is far-distant Munich. And on a wider scale the anomalies multiply. Thus the planning of the Emsland, a thinly populated lowland zone near the Dutch border west of Bremen, needs to be related to the influence of the Ruhr region to the south; but the Emsland is in Niedersachsen and the Ruhrgebiet is in Nordrhein-Westfalen. Most strikingly of all, Berlin's hinterland is all in the surrounding state of Brandenburg, and an attempt to unite the two *Länder* was voted down by the electorate – although joint planning arrangements were then set in place. In France, similar anomalies in the structure of *départements* around Paris were rectified by a stroke of the ministerial pen. But in Germany, it is more difficult.

For good reasons, most of the effort in planning at the national/regional level has been directed at the problems of the remoter rural areas. Despite efforts to maintain agriculture through restructuring, the agricultural population has fallen by more than half since 1950; and it tends to be the younger, more active people who have left, especially from those areas where industrial and other job opportunities were thin on the ground. As it had evolved by the 1970s, the system recognized three main types of area requiring assistance; development areas (*Bundesausbaugebiete*), development centres

(*Bundesausbauorte*) and the frontier zone with the former DDR (*Zonenrandgebiet*). It also recognized four special areas, all of them peripheral: the Emsland against the North Sea in the north west, northern Schleswig-Holstein in the far north, the North Sea coast itself, and the Alps in the far south. In fact all the development areas were themselves originally frontier zones, and a large proportion of them lay against the zonal border with the Democratic Republic; in all they included Schleswig-Holstein, eastern and western Lower Saxony, eastern Hessen and northern Bavaria, the Bavarian Forest (Bayerischer Wald) and the Eifel. All these areas were distinguished by a poor basis in natural resources and a weakly developed infrastructure; they were helped by agricultural reform, tourist development and industry. They included also the depressed coal mining areas of the Saar and of the northern Ruhr.

The development centres, a joint federal–*Land* project, aim to provide the whole country with a well-articulated system and service centres following the principles of central place theory: they are chosen for their suitability in this regard, as well as for the amount of local investment. Most of them have been in the development areas or the frontier zone. Since 1990 the original logic, dictated by the division of Germany, has ceased to apply but the same principles are applied to development in the five new *Länder*.

In all the development areas the policies are similar, and indeed they closely resemble those that have been applied in France. One element is the provision of a better infrastructure, especially in the form of communications; this is important because by definition these areas tend to be away from the main lines of rapid transportation, which run along the major industrial and urban axes. Another is the granting of financial incentives, either as investment allowances or grants, to help private investment, especially in industry. In effect these are now channelled through European structural funds to the five eastern *Länder*, all of which have the strongest level of assistance (Convergence) available, plus the special 20-year programme which Germany developed in 1990 to bring infrastructure in the east up to western levels. The development centres have some similarity in principle to the *métropoles d'équilibre* in France, but in detailed practice they are completely different, because there are many more of them and they tend to be small market towns with some advantages in the form of transport services, cultural and social facilities, and some existing industry, in or near the development areas and so distant from the main currents of economic life (see Plates 7.4a, b). This reflected the fact that in Germany, unlike France, the main regional cities were themselves highly buoyant.

The original programme of aid was only too successful: by the mid-1980s the Federal Ministry for Regional Planning, Building and Urban Development could report that differences in basic infrastructure, and hence in living conditions, had been virtually eliminated throughout the western part of Germany. But by then there was a new priority: aid was more sharply focused on certain areas, to deal with rising unemployment in the older industrial regions and cities, above all in the Ruhr and Saar areas. And then, in the 1990s, came a new challenge in the form of deindustrialization and rising unemployment in the eastern cities. Rapid improvements in infrastructure did attract some major new industrial developments, such as the Volkswagen plant in Eisenach in Harlingen, close to the old border; but generally the balance remained negative.

At regional/local level, planning is once again circumscribed by administrative divisions. The basic unit for physical planning is the municipality, or *Gemeinde*. In much of the country these were formerly very small; but, as a result of reforms in the period 1965–76, all of them now tend to be quite large, so that the urban ones commonly stretch out far beyond the edge of the physically built-up area. Advisory planning associations, called *Landschaftsverbände*, provide some measure of coordination over larger areas, roughly corresponding to city regions. Only in the Ruhr coalfield, from

(a)

Plate 7.4 (a) Main axis of Vauban, a model suburban extension in Freiburg, a university city on the edge of the Black Forest. Since the late 1970s Freiburg has consciously developed itself as one of the most advanced eco-cities of the world, with exceptionally high standards for urban energy consumption and recycling; its public transport system has managed to reverse the growth in car ownership. (b) One of the Vauban neighbourhoods, developed by cooperative building groups within an overall city master plan, producing an exceptionally attractive urban environment with a special stress on shared open space for children.

the early 1920s, was there an effective executive authority exercising real power over a whole urban agglomeration. This unique organization, the Ruhr Coalfield Settlement Association (*Siedlungsverband Ruhrkohlenbezirk*), developed a very imaginative and thorough plan for the region, which involves an attempt to preserve the separate character of the cities which make up the area. By this means, it maintained the limited green spaces which divide one city from its neighbour, and which are so important in limiting the spread of air pollution in this region of 5 million people. Since in general the cities of the Ruhr are aligned on an east–west line – the line of a medieval trade route – and since also the higher land to the south is important as water-gathering grounds and as a recreational area, this means that future growth was encouraged towards the north, where open land is available. Until the decline of the coal industry, this made good sense in terms of economic development, because the zone of main coal extraction was tending to move progressively northwards. And though this is now in doubt, the logic for the emphasis on movement towards the north is irresistible. Apart from anything else, this means an extension of the urban zone and its sphere of influence towards the development areas of the Emsland, so that here regional/local planning arguments are reinforced by national/regional ones.

Plate 7.4 continued (b)

Ironically, in 1977 the SVR was dissolved after more than half a century of successful regional planning. The cause, predictably, was mutual jealousy among the major cities constituting the region. The region did achieve one outstanding success in the late 1980s and 1990s: the Emscherpark, an experimental urban regeneration exercise covering the degraded northern part of the region, carried through as an International Building Exhibition, through no less than 120 different projects, many quite extraordinary: the conversion of a derelict steel works into a site for light and sound performances, or of an old coal mine into a tourist attraction, or the creation of a new east–west park strip as part of 300 square kilometres of new green areas, or a score of highly original experimental buildings (Figure 7.10). But this was a strictly time-limited project that came to an end in 1999, and it remains open to doubt whether the rather frail advisory body that replaced the SVR can maintain its outstanding example.

Nonetheless, Emscherpark represented a landmark in another important respect: some of its model housing schemes were early examples of sustainable urban development, with new standards of construction and new energy sources (in particular solar energy) to reduce dependence on fossil fuel. And during this period and after, Germany progressively developed new global standards in responding to the new challenge of climate change. At national level, in 2001 the country took a major step in establishing the feed-in tariff for energy, whereby ordinary people could generate energy and sell it to the German national grid – a pioneering step which other European countries, in particular the United Kingdom, have been slow to follow. Locally, a few cities have established themselves as global leaders in sustainable urban development. Most notable

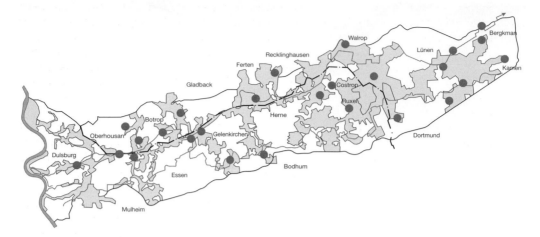

Figure 7.10 The development plan for the Ruhr. The Ruhr regional plan authority pioneered the idea of planning for an entire city region as early as the 1920s. Its 1966 plan (a) aimed to steer development into the open areas in the north of the region, while preserving green barriers between the major cities in the central part. Wide recreation areas would be preserved on the northern and southern peripheries. However, the authority was abolished in 1977 and the strategy from the mid-1980s onward, embodied in the Emscherpark plan (b), is to develop a series of major development projects embodied in a landscape park.

of all is Freiburg, a medium-sized university city on the edge of the Black Forest in south-west Germany, where over a quarter-century a remarkable coalition led by Green Party politicians has worked with the city planner, Wulf Daseking, to achieve new residential areas designed by cooperative building groups, all designed to extraordinarily rigorous standards of energy consumption and recycling. In the early twenty-first century Freiburg has become a place of pilgrimage for planners from all over the world, who come to see how cooperative action can achieve some of Europe's highest standards of planned urban design.

Regional development in Italy

Postwar Italy presents acute problems of planning both at the national/regional scale and the regional/local scale: the contrast between the dynamic industrial economy of the north and the stagnant agricultural society of the south is paralleled by the uncontrolled – and apparently uncontrollable – development of the major city regions, such as Milan–Turin, Rome or Naples. But in all fairness, it must be said that Italy's main innovations, and its main interest for the rest of the world, are at the broad scale of national/regional relationships rather than at the local scale of physical planning controls. As even the most casual visitor to the large Italian cities must notice, the planning machinery does not seem to have been equal to the problems it had to face. Again and again the same features recur: massive traffic congestion due to failure to control the use of the private car in densely built-up cities; a huge stock of obsolescent older housing, seriously deficient in basic facilities; new housing areas which are poorly conceived and poorly located, often without elementary social provision in the form of parks, clinics or shops; and suburban sprawl extending over wide areas, with minimal control over incompatible land uses and spreading traffic congestion. It is true that – especially in the 1990s – efforts have been made to remedy these deficiencies; Milan and Rome have built underground railways, Milan has pioneered the development of priority for bus traffic on the streets, many small cities have banished or controlled the car from their centres and most cities have some attractive suburban areas. And due to the facts of history, solutions are often difficult to find: in few cities can it be said, as it is in Rome, that every few yards the underground railway builder finds a precious historic relic in his path. Nevertheless it is logical that this summary account should concentrate on the larger regional scale.

The centre–periphery contrast is observable in other Western European countries; but nowhere, perhaps, as acutely as in Italy. For hardly anywhere else can it be said that one substantial region of the nation is, in effect, an underdeveloped country; nowhere else is the contrast between the different regions, in the stage and in the speed of their economic development, so great.

The Italian Mezzogiorno was for a half-century after the 1950s the largest problem area in the EU, though for a short time after 1990 the former DDR became in some ways as problematic (Figure 7.11) and since 2004/7 its place has been taken by many regions in the new accession countries. Conventionally defined as the mainland regions of Abruzzo–Molise, Campana and Sud, plus Sicily and Sardinia, it has 20 per cent of the country's area and 20.5 million people, 35 per cent of the population, equal to Belgium, Sweden, Portugal or Greece. Its backwardness has two causes, physical and historical. The area lacks natural resources in the form of coal or hydro-electric power, though oil is found in Sicily; 85 per cent of the land area is mountains and hills, the land is arid and the agricultural possibilities are limited; because of the terrain,

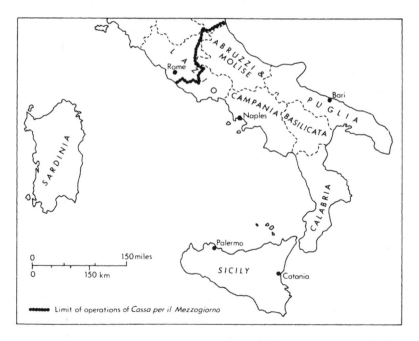

Figure 7.11 The Mezzogiorno. The southern half of Italy is one of the greatest problem regions of Europe. With income levels well below those of northern Italy and a poor economic structure based on subsistence agriculture, this has been a zone of out-migration to the northern cities and the rest of Europe. But the Mezzogiorno has witnessed an ambitious development programme since the Second World War.

communications are poor and to improve them is an expensive job. History has exacerbated the problem: up to unification in the 1860s the kingdom of Naples remained basically feudal, and thereafter the policy of protection encouraged the maintenance of inefficient agriculture at the same time as it permitted the infant industry of the north to grow. The unification of Italy in the nineteenth century hurt rather than helped the region, draining capital and killing its industries. Down to the 1990s, per capita incomes have remained less than 75 per cent of the EU average, half those of the north.

In the post-Second World War period, too, the south's loss has been the north's gain: the constant flow of new labour northwards allowed northern industry to increase its productivity faster than the average wage rate. But the south itself remained massively under-represented in modern industry, especially the critical growth sectors of engineering and chemicals. Most disturbingly, the Mezzogiorno failed to develop concentrations of large-scale industry that could exploit economies of scale and inter-industry linkages. The disequilibrium between Italy's own 'Golden Triangle' (Milan–Turin–Genoa, itself part of the wider European 'Blue Banana'), and the 'Third Italy' of the Emilia–Romagna region around Bologna, vis-à-vis the south, became steadily more marked.

In 1950, partly impelled by fears of civil unrest, the Italian government took a bold initiative: it set up the *Cassa per il Mezzogiorno* (Fund for the Mezzogiorno) for a period of 30 years as a linking operation between its operations and local authorities. In the 1950s it promoted land reform, breaking up the traditional large estates and establishing smallholdings, and supplementing this by investment both in agricultural improvement and in better transportation. But in the course of the 1950s it became apparent that a broader-based strategy was needed, and emphasis shifted to industrial development through industrial credit at low rates (partly through specially created state funds), subsidies for industrial investment in buildings and plants, tax concessions and even the taking of a share in the equity of private firms; more than £1 billion was spent on developing the south between 1955 and 1970.

By the late 1960s these forms of help together represented the highest level of total regional aid available in the EEC, though they were closely limited by EEC policy guidelines, which fixed assistance at a maximum level some 20 per cent above the cost of investment in 'central', highly favourable regions, such as Lombardy. At the end of the 1960s it developed a growth pole strategy, involving a few select sectors – iron and steel, machine building, precision engineering, oil refining and petrochemicals plus linked branches – and four main zones: Casterta–Naples–Salerno, Catania–Siracuse, Cagliari and Bari–Brindisi–Taranto. In the late 1960s and early 1970s the government attracted a wider range of firms, such as Alfa Romeo cars; this was a period of the strongest development, before it as overtaken by the first oil crisis.

The *Cassa* had some success, but critics have pointed to the high operating costs of the plants thus established. Overall the *Cassa* achieved impressive investment: over 35 years, 2,427 factories with 305,000 workers. But these figures exclude job losses which make the overall picture much less impressive. The effort is very concentrated sectorally, with 73 per cent of the new jobs in chemical, metallurgical and mechanical branches, mostly in large, capital-intensive units. Many of these big plants are externally controlled, making the Mezzogiorno a classic case of dependent development; there remains the problem of 'cathedrals in the desert' without local links. Their establishment resulted from the fact that most of the aid consisted of subsidies to investment in an area where labour surpluses were the problem.

Relatively late in the 1960s some attempt was made to correct this by offering remission of social security payments on behalf of their employees to firms establishing themselves in the south – an incentive similar to the regional employment premium introduced in Britain at the same time. But in the 1980s some 40 per cent of the total labour force of the south remained in agriculture, while conversely 40 per cent of the total national industrial labour force was found in the industrial north west. Between 1950 and 2000 the south increased its non-agricultural labour force more slowly than it ought to have done, and manufacturing employment as a percentage of the total was only about two-thirds that in the north; in absolute terms this increase was smaller than the net out-migration from the area, which in most of the Mezzogiorno actually exceeded the above-average birth rate.

Further, the benefits of the policies were highly concentrated in terms of geography: nearly 60 per cent of the jobs were in only four areas: Latina–Frosinone, Naples–Casterta–Salerno, Bari–Brindisi–Taranto and Catania–Siracusa. So industrial development has actually increased disparities. Partly by deliberate intention, the main effects of the programme were seen on the western side of the mainland and in eastern and southern Sicily. As a result the eastern part of the mainland – the so-called heel of Italy – benefited relatively little, apart from the creation of a petro-chemical complex in the Bari–Brindisi–Taranto area. In this policy the planners were concerned to develop linked industries of a more labour-intensive kind, which would naturally associate themselves with the capital-intensive chemical and electrical plants already established in a few of the more developed urban areas of the Mezzogiorno. But it has been argued that such linkages are relatively unimportant in so under-developed an area as the south.

These qualifications point to severe limitations on the effectiveness of the policy. It was not accompanied, as in the United Kingdom, by negative controls on industrial growth in the prosperous north; all that was done was to establish that a certain fixed share of investment by public firms (40 per cent, and 60 per cent of new invest-ment) should be in the south. The generous supply of labour in the south proved to be a minor asset, since rapid migration was producing similar pools in the north, where, also, industrialists enjoyed the advantage of existing industrial infrastructure and complementary industries.

Thus regional policy in Italy has exposed the dilemma of investment priorities. In the growing cities of the north and centre there is an acute demand for investment in new infrastructure; the better economic conditions create the resources necessary for the purpose, especially in the private sector. But at the same time, this programme draws further construction workers from the south; it is done at a high cost, as such urban-structuring always must; and it generates further problems of congestion in the long run. The answer may be a policy of developing counter-magnets in the south, along the lines of the French *métropoles d'équilibre*. Yet, apart from Naples and perhaps Palermo, the Mezzogiorno lacks the existing urban concentrations to provide the basis for such developments. They would have to be painstakingly built up, through the growth pole policy. And the question would remain whether such centres could really spread their effect widely enough to help the bulk of the rural population of the south. As in France and Germany, the question of the size and number of these centres, and their relation to the areas of greatest distress, is a crucial one.

Overall, then, the record of postwar regional planning in Italy is far from encouraging. The south has failed to develop manufacturing industry on any scale, and industrial employment still accounted for between 20 and 25 per cent of the total in 1999; its per capita GDP was still only about half the rest of Italy. What industrial development had occurred was mainly in the capital-intensive sectors such as power, water and oil-refining rather than in modern, labour-intensive industry with capacity to develop linkages with other industry (such as chemicals or engineering). This in turn reflected the predominance of state public utility corporations in the development of the region. There has been a failure to develop large-scale industry. The agricultural policy had largely been a failure, with a remaining agricultural surplus population working for low incomes, with inadequate equipment, on poor land. Unemployment at the end of the 1990s was as high as 20 per cent across most of the region, against 12 per cent nationally or 5 per cent in the industrial north, and total employment had tended to contract over a long period.

Thus the experience of the Mezzogiorno seems to demonstrate that reliance on providing public infrastructure, plus investment by capital-intensive state corporations, is not enough to promote regional development; either the available incentives to private industry to locate in the south were too weak or inertia was too strong, or both. A stronger policy, with some negative controls on industrial growth in the more prosperous regions closer to the heart of the EU, would seem to be necessary. Additionally, the geographical benefits of the development did not spread themselves widely enough through the south; to achieve that, it would have been necessary to couple the wider extension of the public infrastructure (particularly the new motorways) through the region with some planned coordination of private industrial investment. Perhaps the application of French methods of indicative planning, which was beginning at the end of the 1960s, will achieve this by involving private industry and public infrastructure in a common process. Such a policy might achieve the unquestioned main objective of planning in the Mezzogiorno: the provision of new industrial jobs for agricultural workers widely across the whole of the region. In 1984 the *Cassa* went into liquidation. But by then, support for the development of the region came principally through European structural funds, and the entire Mezzogiorno remains classed as an Objective 1 region enjoying the highest level of aid.

Scandinavian city-region planning

At the opposite end of Europe, the main interest for the planner is at the more local scale. Scandinavia as a whole is distinguished by having a remarkable degree of concentration of its population within a few major urban regions; about a quarter of the population of Denmark live in the Greater Copenhagen region, and about 40 per cent of the people of Sweden live in the three main urban areas of Greater Stockholm, Malmö-Hälsingborg and Göteborg. This is a reflection of economic conditions that are almost precisely the reverse of those in southern Italy: agriculture is prosperous and highly capitalized, surplus agricultural labour has long since deserted the land for the cities, and the major urban regions have developed as centres of advanced industry, international trading nodes and seats of administration. This development, however, has been relatively recent; the industrial revolution in these countries occurred much later than in Britain, so that (as in Germany) the development took place in the existing commercial cities rather than in newly developed coalfield towns. Because urban growth came so late – at the end of the nineteenth century and in the twentieth – these cities were able to develop effective town-planning controls almost from the start. Soon after 1900, when it was still a small city of 100,000 people, the city of Stockholm began to buy the land all around, so as to guarantee properly planned development. By 2000, with a population of 1.8 million in Stockholm county (750,000 in the city itself), Stockholm had the most comprehensively planned suburbs in the whole of Western Europe.

Together with postwar Britain, these cities have thus contributed quite disproportionately to modern ideas about planning at the scale of the city region – regional/local planning, as we have called it throughout this book. Soon after Abercrombie's famous 1944 plan for Greater London (Chapter 4) with its emphasis on green belt controls on the growth of the city region and the development of new towns outside it, both Copenhagen and Stockholm produced plans based on different solutions to the problem of urban growth. Just as in London, the available controls were strong enough to guarantee that, in essentials, the plans were implemented.

Specifically, Copenhagen produced its now celebrated Finger Plan in 1948, four years after Abercrombie (Figure 7.12). The scale of the problem was much smaller: against 8 million in the Greater London conurbation and 10 million in the wider region, Greater Copenhagen at that time contained just over 1 million people. But the character of the problem was the same: like London, Copenhagen had grown in annular strips around a single core containing much of the employment, and a radial pattern of roads and public transport routes had reinforced the pattern. (With Copenhagen, however, the city took a semi-circular and not a circular form, because of the existence of a stretch of water – the Øresund – which separates Denmark from the southern Swedish city of Malmö, and was finally bridged in the year 2000.) By 1948 development had reached a critical stage: the outer terminals of the public transport routes were already about 45 minutes from Central Copenhagen, about the same time–distance as the outer suburban rail terminals in the much bigger area of London. Copenhagen could have dealt with the problem of future growth by using the London solution of green belt and new towns; instead its planners decided to increase accessibility to the central city, by new forms of higher-speed transport (suburban railways) along certain preferred axes or fingers, thus extending

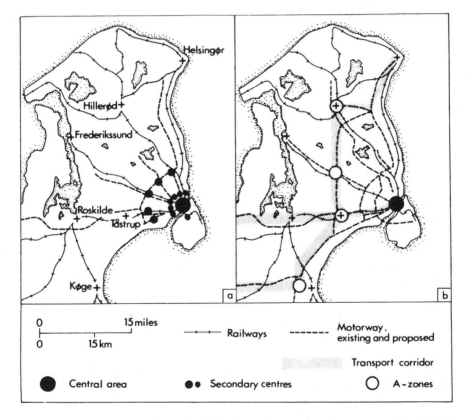

Figure 7.12 Plans for Copenhagen: (a) the 1948 Finger Plan; (b) the 1973 Regional Plan. Copenhagen reacted early against the green belt ideas of the 1944 London plan, substituting a design based on fingers of urban development with intervening green fingers. The later plan extends the fingers with major new employment centres at their junction with a new transportation axis, relieving pressure on the city.

the 45-minute zone much farther from the city centre. Between these axes, wedges of open space would naturally be preserved in the lower-accessibility areas.

The plan was implemented, but Copenhagen – like many other European city regions – grew faster than had been expected; by 1960 the population had already reached 1.5 million, the long-term figure in the 1948 Finger Plan. The revised estimates showed that in the absence of comprehensive controls on industrial location, such as those in Britain, the total could swell further to 2.5 million by the end of the century; and rising space standards would create additional demand for land. With a city growing at such a rate, it was no longer possible to think merely of increasing accessibility to the centre; as in the Abercrombie London plan, jobs, too, must now decentralize. But the Copenhagen planners continued to reject the principle of the green belt and self-contained new town. Instead, they proposed new 'city sections', or major centres, developed on further extensions of the fingers; these would contain both manufacturing and service jobs, and the level of urban services would be appropriate to the average size of each major centre: approximately 250,000, the size of the largest provincial cities of Denmark. Thus jobs for many of the new residents would be provided near home; but for those who must still commute to the centre, very high-speed transport links along the fingers would be available. Overall, it was calculated by the Copenhagen Regional Planning Office that savings in travel costs for such a decentralized structure could be of the order of £50 million per annum.

A split developed over how the principle should be implemented. The four necessary new city sections could be grouped in one sector, with a concentration of the transport investments in that sector; in that case, the logical sector to choose would be that running westwards towards the town of Roskilde and south-westwards towards the town of Køge, both medium-sized country towns about 20 miles (35 kilometres) from the centre of Copenhagen. Alternatively, they could be developed in a number of different sectors, in order to relieve overloading in any one of them; thus, while the first development would take place towards the west and south west, later emphasis would shift to the north west. But the latter development would invade high-quality landscape, important for recreation, which many planners thought should be conserved. Even deeper down, there was a division between those who wanted to encourage rapid decentralization of people and jobs from the city of Copenhagen, and those who wanted to encourage replanning and redevelopment within the city; this, of course, was as much a political as a professional controversy. As a provisional step, development during the 1960s was concentrated in the south-west area, with the development of a new centre next to the suburban railway station at Taastrup, about 12 miles from the centre of the city. By 1971, to provide for further growth, the regional planning council was providing alternative sketches based on concentration along main transportation corridors – not merely the radial lines, but also a north–south line bypassing the main urban agglomeration on the west side.

During the 1970s Copenhagen moved in steady stages, including a public debate as to alternatives, towards a further development of the regional plan. The final choice was based on a new, long-term system of 'transportation corridors', one running north–south from the narrow sea crossing to Sweden at Helsingor (Elsinore) towards Germany, the other running east–west from Copenhagen towards the western region of Denmark. Along them, reservations were made for employment centres, and where they intersect with the radial routes that provided the basis for the Finger Plan, their new regional sub-centres can be established. However, this would not occur at every such point; after debate, the decision was made to concentrate development in four nodal points. Three, west of the city, would help complete the urban structure in that sector. A fourth, south-west of the city, would complement the 1960 plan by serving as a bridgehead for further expansion towards the south. In contrast, growth in the attractive northern open zone would continue to be restricted. This 1973 Regional Plan was confirmed in 1975 and implemented during the following decades, as the regional population grew to 1.8 million at the century's end. Major growth took place around the rail-bus transport interchange at Høje Taastrup, 12 miles (20 kilometres) west of central Copenhagen, though it was somewhat compromised by construction of a shopping mall which has inhibited commercial development in the central core. Interestingly, the plan called for the construction of some quarter of a million new dwellings, even though, as elsewhere in Europe, the regional prognosis is for a stable population. The explanation is the need for urban renewal plus new household formation. And in the 1990s, there was a marked shift in emphasis to regeneration along the former docks near the city centre, and construction of a linear 'new town intown', Ørestad, along a north–south corridor between the city centre and the airport, crossed by the rail link to the new bridge crossing.

In Stockholm the story has been a simpler one – partly, perhaps, because until recently most of the new development took place within the city limits and on land actually owned by the city. An important fact is that the Stockholm agglomeration was smaller than the equivalent Copenhagen one, but has caught up: while Copenhagen reached 1.1 million people by 1945, 1.5 million by 1960 and 1.8 million by 2000, the corresponding figures for Stockholm are 850,000 in 1945, 1.2 million by 1960 and 1.8 million by 2000. In fact, by the mid-1940s – when plans for comprehensive future development were under discussion – Stockholm had already spread to an average distance of about 8 miles (13 kilometres) from the centre, mainly on the basis of tramways; in certain directions, where suburban railway service was available, it extended farther. But in general, the city could not extend much farther on the basis of existing transport systems.

This was one critical fact in the plans produced in the late 1940s, which were embodied in a general plan for the city in 1952. While still a relatively small city by European standards, Stockholm determined that its future growth should be based on an underground railway system consisting of lines radiating from a central interchange station in the city centre, with stations at approximately half-mile intervals. Then, to ensure that the improved accessibility would not be followed by low-density suburban spread, as in London during the 1920s and 1930s, new suburbs would be deliberately planned around the new railway stations on the principle of local pyramids of density: higher densities around the stations, lower farther away. The railway station areas were also logically to become centres for shopping and other services, ordered according to a hierarchical principle; thus, the new shops would have the maximum number of customers easily placed within walking distance, in the surrounding high-density flats. Lastly, the system of suburban areas thus created would be physically defined by local green belts, which would wrap around and interpenetrate them (Figure 7.13, Plate 7.5).

These principles were faithfully followed in the rapid urban development of the 1950s and 1960s. By the mid-1960s the underground system extended over a 40-mile network, serving the new suburban areas, and bringing them all within a 40-minute (64-kilometre) ride of the city centre. At the railway stations, shopping centres were built at one of two levels in the hierarchy: local 'C' centres serving 10,000–15,000 people, mainly within walking distance of a station, and sub-regional 'B' centres serving several suburban areas, with 15,000–30,000 people within walking distance and another 50,000–100,000 served by underground, feeder bus or private car. To meet the rising demand for shopping by car, parking provision at the 'B' centres has sharply increased since the original design of the mid-1950s; and these centres are served by high-capacity arterial highways, which were upgraded to motorway standard in the late 1960s and early 1970s. Thus, in a typical suburban group, four local 'C' centres would be grouped around their 'B' centre, the whole being tied together both by the underground line (running usually above ground in the suburbs) and by the highway system.

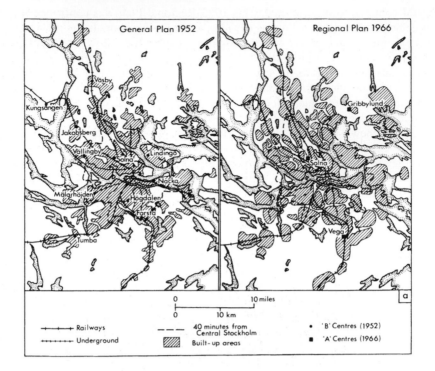

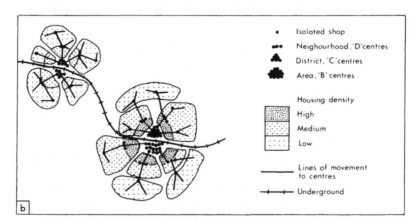

Figure 7.13 (a) Plans for Stockholm: 1952 and 1966. The 1952 plan established the idea of planned suburban satellites, with a hierarchy of shopping centres, linked by the new underground railway system. This plan was largely implemented by the late 1960s, when a wide-ranging regional plan extended the principle through new developments along main-line railways and motorways radiating from the city. (b) Stockholm: the schematic diagram of a suburban group – the principle of the hierarchy of suburban shopping centres in the 1952 plan. The bigger 'B' centres are surrounded by high-density residential areas from which the inhabitants can walk to the shops or to the underground station. Densities fall away from the centre towards the edge.

Plate 7.5 Aerial view of Farsta, Sweden: the centre of one of the planned Stockholm suburbs, designed in the early 1960s. The underground station is seen in the right-centre of the picture, surrounded by higher-density residential developments. The shopping centre is easily accessible by car, underground train or on foot from the apartment blocks.

Together with the British new towns, the Stockholm suburbs represent one of the most admired planning achievements of the mid-twentieth century. But by the late 1960s, limitations were apparent. One was that unlike the British new towns policy, the 1952 plan did not ensure large-scale decentralization of jobs; indeed, in the early 1960s the central area (Hötorget), over and around the central underground station, was reconstructed to provide for big increase in office jobs and shopping, while the new suburban areas failed to become self-contained towns for working and living on the British model. In a relatively small city with exceptionally good public transport, this ideal was probably unattainable and undesirable. But as the size of the developed area grew, a new scale of thinking was necessary. This is the basis of the Greater Stockholm regions plans of 1973 and 1978 – plans which take in not merely the city, but also adjacent suburban and rural areas up to 20 miles (30 kilometres) in all directions.

These plans start from certain principles, which emerge from research studies and forecasts. One was that though space-using types of industry would seek peripheral locations, an increasing number of decision-making service industry jobs would still seek locations at or near the centre. Another is that people would seek more space in and around their homes, leading to a big extension of the total developed area.

The 1978 version of the plan assumed that by 1990 the inner city, with 13 per cent of the population, would still have 32 per cent of the workplaces. The other main work area, the north-west sector, would have 29 per cent of people and 25 per cent of employment. The remaining sectors, especially the south east, would be relatively short of job opportunities. Thus, despite a conscious attempt to plan homes and jobs in close proximity, the result would be an increasing demand for long-distance commuter journeys to the centre; and since the structure of central Stockholm will not allow for more than a small proportion of these to be made by private car, major investment in rail transport is a priority. During the 1970s the underground system was further extended to serve a major new development for a total of 32,000 people in the Jarvafältet area, north-west of the city. Because, however, it is too slow to serve effectively those areas more than about 12 miles from the centre (also the effective radius of the London system), future growth outside these limits must depend on faster long-distance commuter services on the main-line Swedish railways system. In this way, it was hoped to secure an actual reduction in the percentage or really long (45-minute or more) commuter journeys by public transport. The new suburbs themselves would not concentrate so closely around the stations, but would take a more dispersed form, following feeder bus routes. Thus the future city region would tend to take a star-shaped form, with long fingers of development following the main-line railways and parallel national motorways, westwards to Södertälje, northwards to Arlanda airport and south-eastwards to Tungelsta. Along these axial extensions, groups of neighbourhoods would constitute physical units, separated from each other by belts of open land that contain the major highways for longer-distance movement. Overall the urban structure would be rather discontinuous, with large areas of open land – which would be heavily used for summer homes and recreation – in contrast to the compact nineteenth-century town (Plate 7.6).

During the 1990s both Copenhagen and Stockholm pioneered a new concept in transport planning as a basis for long-term development: the regional metro. This consisted in linking up longer-distance train lines of their respective national railways so as to provide a new level of express commuter services as well as specialized services connecting to their respective airports. In Stockholm, the 1991 Mälardalen Regional Plan proposed the use of new high-speed train services as a way of linking Stockholm with cities in this distance range such as Ensköping, Västerås, Eskilstuna and Örebrö. This was the first case in which a regional development plan was deliberately structured around the existence of high-speed links. In Copenhagen, the Øresund link between Sweden and Denmark, opened in 2000, carries the world's first international regional metro connecting the Danish cities of Roskilde and Køge via central Copenhagen and Copenhagen Airport with the Swedish cities of Lund and Malmö – and, as seen, this connects planned new communities outside Copenhagen with Ørestad, the new town intown close to Copenhagen's airport and with the vast Västra Hamnen (Western Harbour) regeneration project in Malmö.

This scheme, a model of the new Swedish emphasis on sustainable urbanism, interestingly illustrates a new trend in Scandinavian planning: as elsewhere in Europe (Ørestad in Copenhagen, Seine-St-Denis in Paris, Amsterdam Harbour in the Netherlands), the focus has shifted sharply towards urban regeneration projects on abandoned inner-city industrial land. These are based on moderately high-density apartment living close to the city centre, served by excellent public transport, with very high design quality (often, as here, with a waterfront location), and with a major emphasis on energy consumption and waste management both within and outside the individual units. A closely comparable development in Stockholm is Hammarby Sjöstad, a 160-hectare site 2 miles (4 kilometres) south of the city centre, planned for an eventual total of 10,800 apartments, at an average density of 100 residential units per hectare in the residential areas (67.5 units per hectare overall) in a superb waterside setting. Developments like this are attracting professional workers back into the inner city, including families with children – thus surprising the planners, who had not expected it. It may well mark a trend: as the affluent reoccupy the city, some at least of the famous satellite towns are becoming the homes of lower-income immigrant workers and their children. The proud boast of the planners of the 1950s and 1960s – that all Swedes lived together, regardless of class – is increasingly being eroded.

Plate 7.6 Hammarby Sjöstad, Stockholm. A model eco-development along the waterfront in inner Stockholm, on the site of an old industrial area and served by a new orbital tramway line, Hammarby consists of moderately high-density apartment blocks with attractive public open space. There is particular stress on low-energy construction and on an ambitious recycling system through vacuum tubes to a central station. Designed for singles and childless couples, it has paradoxically proved attractive to families with young children. Unlike the 1960s suburbs it is, however, a commercial development for an affluent market.

The Netherlands: Randstad and regional development

For the last case study in this chapter, we return southwards to the heart of the EU. The western Netherlands sit in the centre of that smaller Golden Triangle, located near the north-western heart of the EU area, in the heart both of the Blue Banana and the Pentagon, where such a high proportion of the economic life and urban population are concentrated. In the great port and industrial complex at the Rhine mouth (Rijnmond), the Dutch may fairly claim to have the principal point of exchange between the EU and the rest of the world. By any reasonable standards of international comparison, here is one of the most important city regions of the European Continent, with a population close to 6 million (Figure 7.14, Plate 7.7).

It is, however, a city region of an unusual kind. No less than 36 per cent of the population of the Netherlands is concentrated here, on 5 per cent of the land area – a degree of metropolitan concentration greater even than in Britain or in France. Yet the impression on the observer is certainly not one of metropolitan over-growth. This is because the Dutch metropolis, which the Dutch themselves call Randstad (or Ring City) Holland, takes the form of a ring of physically separate cities, running in an approximately horseshoe-shaped line approximately 110 miles (190 kilometres) in length. At the start of the twenty-first century it incorporated three big cities with a combined population of over 1.8 million – Amsterdam (762,000), Rotterdam (585,000) and the Hague (486,000) – as well as four other important cities: Utrecht (302,000), Almere (187,000), Haarlem (149,000) and Leiden (118,000). Each is separated from its neighbours by a green zone, even though this is sometimes wafer-thin in the extreme western part of the horseshoe. All the cities look inwards into a central area of open space, carefully preserved by regional planning, which has earned the whole complex the nickname (in the words of the British planner Gerald Burke) of Greenheart Metropolis.

Unlike London or Paris or Stockholm, but like the Ruhr area of Federal Germany, the Randstad is therefore an example of a polycentric metropolis. This quality is not merely physical; it is also functional, in that different cities within the complex perform

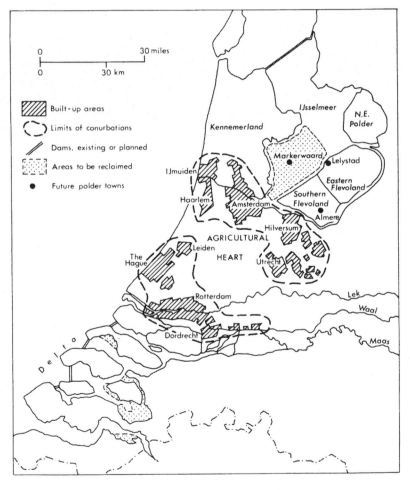

Figure 7.14 Map of Randstad Holland. Urban development in the western Netherlands has taken the form of a horseshoe shaped ring of cities, each performing specialized functions (government in The Hague, commerce in Rotterdam, shopping and culture in Amsterdam), with a central 'green heart' which it is planning policy to protect. The Markerwaard reclamation has not been completed.

Plate 7.7 View of Randstad Holland, near Rotterdam. The mixture of new housing and intensive agriculture is typical of the Randstad – a polynuclear urban region in which town and countryside rapidly alternate.

broadly different functions. Thus government is concentrated in The Hague; the port, wholesale business and heavy industry in Rotterdam; finance, retailing, tourism and culture in Amsterdam; lighter manufacturing and more local service-provision in a number of smaller centres. By splitting functions into separate cities in this way, the Randstad avoids several of the more grievous problems of larger single-centred metropolitan cities: journeys to work tend to be shorter, traffic congestion less widespread. But there is one problem that the Netherlands shares with its EU neighbours: it is the problem of regional imbalance between the booming growth of the Randstad and the more laggardly development (or stagnation) of peripheral areas such as Limburg (in the southern part of the country) or Groningen (in the north west, on the far side of the former Zuiderzee). Thus there have developed ambitious programmes of regional development, in the form of incentives to locate in these regions, coupled with the provision of infrastructure; but as is usual throughout the EU, in contrast to Britain down to the 1980s, this did not extend to an attempt to limit economic growth in the western Netherlands by actual restraints on new factories or other industry.

By the 1960s it appeared that the policy of encouraging the peripheral regions was having an effect; the proportion of national population growth in the western provinces was falling, though much of the increase was passing to immediately adjacent regions. But because of the very rapid natural growth of the whole population, this still meant a very rapid increase in the Randstad itself. Therefore, in order to encourage further decentralization, the government physical planning service proposed that the Randstad should grow outwards along lines of good accessibility, especially into areas where land reclamation or new lines of communication, or both, create new opportunities. Outstanding among these were the reclaimed polders of the former Zuiderzee – where, 40 years later, Lelystad, the central city, had a 2008 population of 73,000 (some 30,000 below its original planned target) and Almere, started later but better placed because closer to Amsterdam, had 187,000 – and the delta region south of Rotterdam. Other opportunities for growth would occur on older reclaimed land, in the northern tip of the province of North Holland adjacent to the west end of the Enclosing Dyke across the former Zuiderzee. The Randstad, thus extended, was expected to contain no less than ten concentrations of 250,000 and more people by the year 2000. The critical factor, therefore, was the way population and employment are distributed within these major concentrations, or city regions; and this was a main theme of the Second Report on Physical Planning in the Netherlands, published in 1966.

This report started from several theses about present trends and future projections. Jobs were expected to decentralize somewhat from city centres, and even out of the Randstad altogether. Residential areas would spread out even more, because a majority of the population – 50–70 per cent – would wish to live in single-family homes. Densities could be lower in smaller urban units, but these would offer less variety of urban services. Whereas the smallest towns could accommodate fairly widespread car usage, this became progressively more difficult as the urban size increases; and above about 1–2 million in size, the problems would increasingly demand rail-based systems for their solution. The urban structure which emerged from this analysis is based on a hierarchy of differently sized units, which could theoretically be combined in different ways: these ranged from a local unit for about 5,000 people, through a unit of about 15,000 and another of 60,000 to one of 250,000 which could offer a very complete range of urban services. (The two middle levels of this hierarchy seem similar in many respects to the 'C' and 'B' levels of the Stockholm planners' hierarchy.) The problem was how these units were to be combined on the ground.

The preferred solution, in the Second Report, was termed 'concentrated deconcentration' (Figure 7.15a). Essentially, this was a compromise between the two extremes of concentration – which would give high accessibility to jobs and services, but poor environment for living – and deconcentration, which would use too much space. The preferred solution would offer a good choice in terms of job opportunities, housing patterns, modes of transport and types of recreation; and it was flexible, since it does not put a rigid shape on the future development of the city region. Units at each level of the hierarchy were separated, but were grouped closely together. In the biggest clusters (such as Amsterdam) all four levels were represented; in smaller ones (such as Haarlem) the topmost regional level was missing, but was available not far away. About one quarter of the population would live in the smallest units and about one quarter in the largest; the average housing density would be four times as great in the latter as in the former (about 60 dwellings per hectare against 15). Applied to the Randstad, the scheme gave six top-level units (based on Dordrecht, Rotterdam, The Hague, Amsterdam, Utrecht and Arnhem), and some 40 centres at the next level some of them independent, some in the form of suburbs attached to the bigger centres. The two lower levels of the hierarchy formed either outer suburban centres for the biggest city regions, or independent villages and small town centres, or systems linking these regions along major routeways. The central green heart of the Randstad would be preserved to a remarkable degree, with only a few urban centres of the lowest status; all the emphasis was on outward development northwards, north-eastwards, eastwards and southwards. But it would be a mistake to treat this as another version of the axial plans of Stockholm or Copenhagen (or, for that matter, Paris); the future urban structure of the extended Randstad, based

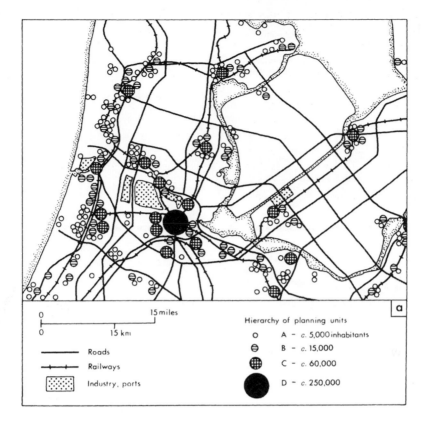

Figure 7.15 (a) The principle of 'concentrated deconcentration', from the *Second Report on Physical Planning in the Netherlands*. For the future development of the Dutch Randstad, government planners suggest grouping the population into urban agglomerations which could then be allowed to decentralize to give a variety of living conditions; urban, suburban and semi-rural. (b) The 'ABC' principle, from the *Fourth Report Extra*. Employment centres are classified in three types. 'A' locations are dense concentrations in city centres, served by excellent public transport. 'B' centres, at city edges, have both rail and highway access. 'C' locations, adjacent to motorway interchanges, are for space-consuming activities like warehousing, which employ relatively few people.

on the careful articulation of a number of differently sized building blocks, is a very complex one which has no real parallel elsewhere (Plate 7.8a, b).

The 1966 plan, in reality, suffered from failures in implementation, but for two strangely contradictory reasons. First, the relatively small municipalities of the Randstad proved too weak to resist pressures for development in the green heart. But second, sharply reducing birth rates during the 1970s worked, together with pressures for out-migration, to produce a new phenomenon: loss of population from the Randstad towards the southern and eastern parts of the Netherlands, associated with the growth of long-distance commuting as these new suburbanites returned to their jobs in the Randstad cities. This, of course, flew in the face of reality in the energy-conscious mid-1970s. Therefore, the Third Report on Physical Planning, in 1974–6, recognized these new facts by abandoning the aim of planned long-distance decentralization from the Randstad and substituting a new urban structure inside the area, based on rather higher densities in rehabilitated inner cities and a more spread-out population elsewhere; there would be growth centres close to the major cities, and extensive upgrading of the existing

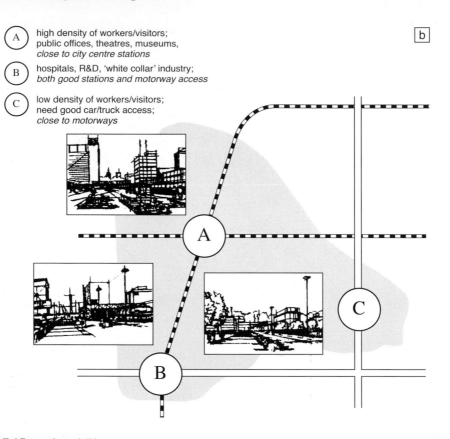

(A) high density of workers/visitors;
public offices, theatres, museums,
close to city centre stations

(B) hospitals, R&D, 'white collar' industry;
both good stations and motorway access

(C) low density of workers/visitors;
need good car/truck access;
close to motorways

b

Figure 7.15 continued (b)

stock in the cities themselves. Both these cities, and the growth centres nearby, would be planned as integral city-region wholes, and the green heart of the Randstad would be resolutely guarded. This, the government recognized, might mean new administrative arrangements, including the creation of larger, stronger municipalities. Down to the early 1990s, little has been done to bring this about. Nevertheless, the Dutch remain as committed as ever to the idea of comprehensive, integrated regional and urban planning for one of the world's most complex urban areas.

In fact, during the 1970s the four major cities lost as much as 15 per cent of their populations, and by the 1990s they had become quite different in important ways from the rest of the Netherlands: they had lost significant numbers of industrial and port jobs, they were suffering much higher levels of unemployment and deprivation, and they had much higher proportions of ethnic minority populations. Further, in the so-called southern wing of the Randstad, Rotterdam was demonstrating similar features – port closure, industrial decline, employment loss, land abandonment – as older industrial cities elsewhere in Europe. Nevertheless, just because economic change had created significant areas of vacant urban land, like their British counterparts they experienced very significant regeneration in the 1990s, leading again to sharp increases in population. Reclaimed dockland areas in Amsterdam (Java Island) and Rotterdam (Kop van Zuid) have been comprehensively rebuilt, and the results are so spectacular that they have attracted flows of professional visitors from other countries, anxious to see how the Dutch have managed the job.

Plate 7.8 (a) Vathorst: aerial view of a planned urban extension on the edge of Amersfoort, a city at the periphery of the Randstad that is served by a new train station and dedicated BRT (Bus Rapid Transit) system along its central axial boulevard. (b) A Vathorst neighbourhood: part of the VINEX housing programme, this illustrates the extremely high quality of Dutch urban design, which compares with the best recent German and Swedish examples.

But there is a remaining doubt. In 1991 a supplement to the Fourth Planning Report (the Fourth Report Extra, in Dutch *Vierde Nota Ruimtelijke Ordening Extra* or VINEX), had developed an important policy prescription that evoked interest and even imitation elsewhere: all employment centres should be classed into three categories (Figure 7.15b). 'A' centres, in the hearts of the cities, had excellent access by public transport from places near and far, though car access was limited and should be further limited: they should be the main concentrations of dense employment. 'B' locations, on the city fringes, enjoyed less spectacular but still adequate access by rail or tram, but also good access by car via radial and orbital motorways; they were suitable for secondary concentrations that needed more space, for instance exhibition and conference centres or stadia. 'C' locations required a lot of space for activities like freight logistics, but needed few employees; they alone should be allowed and even encouraged to locate near motorway interchanges, away from rail lines. The distinction was fine – in theory. But in Amsterdam developers avoided the 'A' location around the central train station, apart from waterfront residential apartments, preferring to put commercial development next to the South Station close to Amsterdam's Schiphol airport, a classic 'B' location, and eventually planning policies accepted the fact, developing the concept of a new 6-mile (10-kilometre) linear central business district along the ring motorway and parallel rail line from Amsterdam's new arena to the airport, the *Zuidas* (southern axis): the

Plate 7.8 continued (b)

Dutch may be very hands-on planners but they are also realistic and pragmatic, and what is good for business is finally good for the Netherlands. Here, perhaps, is a lesson that British planners might usefully learn.

The VINEX report was important in another sense: it outlined a vision for a ten-year crash housing programme to compensate for accumulated shortages. Local authorities were the agents: the central government provided funding to decontaminate land and provide access, but otherwise the schemes were self-funding. It has been phenomenally successful, producing no less than 90 schemes and 450,000 new homes, and increasing the total Dutch housing stock by 7.6 per cent. Further, the homes have been built close to existing cities in order to minimize invasion of valuable greenfield land, above all in the 'green heart', to minimize travel to the cities and to secure maximum use of public transport, walking and bicycles. There is also an element of social engineering: VINEX aimed to lure higher-income households out to the new locations, thus freeing up housing in the cities for lower-income residents – but also providing a share of cheaper rented housing in the VINEX locations themselves.

Among the 90 locations there were 25 major schemes, located all over the country but with a marked preponderance in the Randstad. Nearly two in five of the new units have been built within city boundaries, some in major urban regeneration schemes like IJburg in the Amsterdam docks, but sometimes – as in Vathorst and Nieuwland next to Amersfoort – in urban extensions right on those boundaries. All are, however, close to the centre of the nearest major city and well connected to it by good public transport – typically with journey times of half an hour or less. And some – like IJburg, a brownfield regeneration scheme in the Amsterdam docks, or Vathorst at the edge of Amersfoort, or Ypenburg next to The Hague – are among the best examples of European urban design in the early twenty-first century.

Bringing Europe together: the European spatial planning perspective

To generalize from such diverse cases may seem impossible. But in their different ways, the countries of the European mainland do illustrate some common points.

The first concerns national/regional planning and the centre–periphery contrast which, in one form or another, recurs in all these countries. (Though it was not treated for lack of space, it can be found too in Scandinavia, in the problems of development of such areas as northern Jutland and the northern half of Sweden.) A real danger, which concerns many European planners, is that the EU may actually reinforce, rather than diminish, this imbalance. The EU-27 countries are clearly tied together by strong trade lines which connect up their major urban areas – lines like the Rhine and the more important Alpine passes, or the Rhône–Saône corridor plus the Mediterranean coast of France and northern Italy. The increased economic links between these urban areas, along the above lines, may have the effect of making the peripheral areas seem even more remote. And technological advances, like new motorways or high-speed railways along these lines, may exacerbate this relative 'peripheralization of the periphery'. Moreover, by increasing economic opportunities in the urban areas, they may accelerate the process of rural depopulation. This was the experience in France and Germany in the 1960s, repeated in Eastern Europe in the 2000s.

Against this, the peripheral areas may be able to offer low labour costs – an advantage that may become more telling, since with inflation labour costs tend to become a steadily larger proportion of total costs. This has been the experience in Eastern European countries like Slovakia, which have attracted new manufacturing plants even before full accession in 2004.

A second point about the national/regional scale concerns the measures which European countries have taken to deal with the problem. Overall, with the exception of the French controls on new establishments in the Paris region, these countries have conspicuously avoided the sort of negative controls which the British operated for industrial development between 1945 and 1982 and for office development during 1964–5, but are now a mere memory there too. They have relied heavily on inducements, generally in the form of grants and loans for building or equipment of new industry in the development areas, coupled with provision of state infrastructure, especially in the form of improved communications with the outside world. To varying extents they have also operated a policy of trying to channel aid into cities or towns which seemed to be favourably located within the development areas – though the expression of this policy varied very greatly, from the giant *métropoles d'équilibre* of France to the much more modest *Bundesausbauorte* of Germany's remoter rural areas. The policies have met with mixed success, but in general it cannot be said that the results have been spectacular.

Here, however, a word of reservation is necessary: both the problems at the regional/local scale and the solutions are necessarily rather different from those in postwar Britain. The problem is one of agricultural depopulation rather than decline of industrial staples; the solution has been to encourage industry to move into the countryside rather than to build up new industry to replace the old. Only in recent years, in areas such as Germany's Ruhr area and its eastern German cities, in the French northern coalfield or on Spain's Basque coast, have problems of industrial adaptation arisen on the mainland of Western Europe.

A significant point about the EU, however, is how slow it and its predecessors were to develop a common regional policy. The Treaty of Rome allows the EU Commission to challenge policies which distort competition. But this cannot affect a national policy which is non-discriminatory with regard to the national origin of the product and which does not distort national competition. Because of this, as we have seen, different European countries have vigorously pursued various forms of aid to industry without interference. By the 1970s, there were firm Commission recommendations as to the extent of aid to industries in needy regions outside a central zone. These guidelines seek to prevent any permanent subsidy to industry in depressed regions and they provided one potent reason why, from 1976, Britain was forced to dismantle its cherished Regional Employment Premium. The EU, however, does temper these controls with regional incentive; the European Investment Bank, with low-interest loans to cover part of the costs of modernization for new enterprise; special loans for the development of new activities in depressed coal and steel areas; and potentially most important, the Regional Fund, which has been viewed especially to attract more labour-intensive activities to the depressed regions.

On the regional/local scale, all these countries have faced the problem of the continued growth of large metropolitan areas. Though there are few parallels elsewhere in Europe to the scale of problem represented by London or Paris – Madrid, which has grown rapidly into a metropolis of 6 million people, is the sole exception – the solutions adopted for smaller-scale metropolitan cities, such as Copenhagen or Stockholm, may prove apposite for many other cities of similar size in other countries. Most significantly, perhaps, the experience of polycentric metropolitan areas, such as the Dutch Randstad or the Rhine–Ruhr region, provides some possible object-lessons for the future internal organization of very large city regions. Certainly, insofar as comparisons can ever be meaningful between such individual and varied urban areas, these polycentric urban regions do seem to avoid some of the more acute problems that afflict their monocentric equivalents, such as London, Paris or New York.

The question is whether it is now possible to develop a common approach on principles that would be broadly acceptable across the European space, by national member governments which would have to bear the main burden of implementation. It may be possible: the EU's member states share many common problems, common concerns and broadly similar policy objectives. All are concerned to promote economic development in order to generate new jobs and new wealth to replace the losses that arise from globalization processes, especially deindustrialization. All wish to promote social cohesion so as to reduce the problems that arise from concentrated deprivation and social exclusion. All, driven by the threat of climate change, subscribe to the principle of sustainability and wish to apply it to urban development. All want to promote a more balanced distribution of economic development across their national territories, so as to reduce regional disparities. Of course, the resulting policy prescriptions are likely to have a different emphasis in different regions of the EU: in some they may suggest the need to regenerate older industrial cities, in others the desirability of enhancing urban infrastructure in smaller rural service towns. But the principle remains the same.

The central question, then, is how these objectives translate into policies and how then those policies receive a spatial dimension. The European Spatial Development Perspective (ESDP), nearly ten years in formulation and finalized in 1999, adopts a central principle: *polycentricity*, allied to *decentralized concentration*: a principle, noticed earlier in this chapter, of Dutch spatial planning, which aims to disperse economic development from congested urban regions in the central Pentagon – a process that is already happening (Figure 7.16) – but to reconcentrate it in urban centres in the less developed regions, thus benefiting both kinds of region.

However, the ESDP does so at the largest possible geographical scale. The aim is less to redistribute some fixed amount of activity in a kind of zero-sum game, than to encourage a significantly higher level of growth in less-developed regions and cities: some older industrial cities in need of restructuring, many others cities in the less densely populated, less-developed fringe regions of Western, Southern, Northern and Eastern Europe.

Thus, the central word, *polycentric*, needs to be carefully defined: it has a different significance at different spatial scales and in different geographical contexts. At the global level, *polycentric* refers to the development of alternative global centres of power. Presently, there are a very few cities worldwide that are universally regarded as global control-and-command centres, located in the most advanced economies: London appears in all lists, Paris appears on some. But Europe has a number of 'sub-global' cities, performing some global functions in specialized fields: Rome (culture), Milan (fashion), Frankfurt and Zurich (banking), Brussels, Luxembourg, Paris, Rome and Geneva (supranational government agencies). Within a European context, therefore, one meaning of a *polycentric* policy is to divert some activities away from 'global' cities like London (and perhaps Paris) to 'sub-global' centres like Brussels, Frankfurt or Milan. But there is also a very important spatial dimension: while some of these cities are found in the European core region (Brussels, Amsterdam, Frankfurt, Luxembourg), a much larger number are 'gateway' national political or commercial capitals outside the centre capitals region: Helsinki, Stockholm, Copenhagen, Berlin, Vienna, Rome/Milan, Madrid/Barcelona, Lisbon and Dublin. They serve broad but sometimes thinly populated territories such as the Iberian peninsula, Scandinavia and east central Europe. Because they are national capitals serving distinct linguistic groups, they invariably have a level of service functions larger than would be expected on grounds of size alone; they tend to be national airport and rail hubs, and the main centres for national cultural institutions and national media.

A major issue here is whether it will be either necessary or desirable to concentrate decentralized activity into a limited number of these 'regional capitals', each commanding a significant sector of the European territory – Copenhagen, Berlin, Rome, Madrid – or whether it would be preferable to diffuse down to the level of the national capital cities, including the smaller national capitals. Essentially, how far should Madrid be regarded as the dominant gateway for south west Europe, or should it share this role with Lisbon, Bilbao, Barcelona and Seville? And likewise with Copenhagen *vis-à-vis* Stockholm, Oslo and Helsinki? This could be particularly important in Eastern Europe, where Berlin and Vienna may develop important roles for their hinterlands reflecting past geographies, but where also there is a real need to reassert the service roles of the different national capitals and selected provincial capitals (Gdańsk, Cracow, Pilsen, Szeged).

But at a finer geographical scale, polycentricity can refer to the outward diffusion from either of these levels of city to smaller cities within their urban fields or spheres of influence. In Chapter 6 we saw that such a process has occurred widely around London, producing a polycentric 'Mega-City Region' extending up to 80 miles (130 kilometres) from the capital; around Paris and Berlin this is much less evident (Figure 7.16 shows urban spheres of influence in the EU, from a 1997 draft of the ESDP that was not reproduced in the final version). At the next level, cities like Stockholm, Copenhagen and Milan show widespread outward diffusion while other cities do not. East European cities, in particular, have had relatively little impact through decentralization on their surrounding regions, though this may change in the future.

In general, at this scale a policy of 'deconcentrated concentration' would suggest adopting the principle fairly widely, but adapting it to the specific development stages and problems of each city and region. The principle should be to guide decentralized growth, wherever possible, on to a few selected development corridors along strong

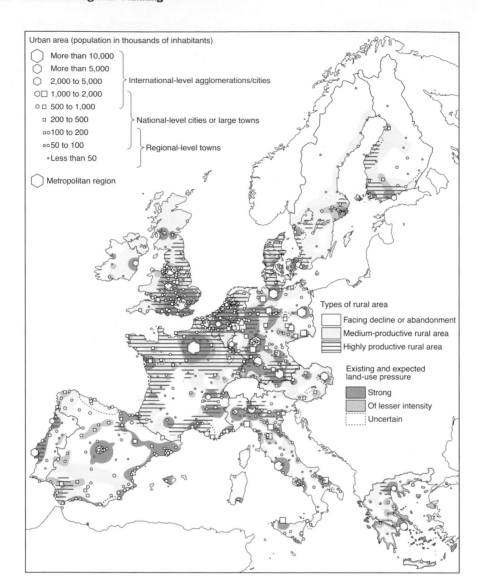

Figure 7.16 Urban spheres of influence in the EU: a map from a draft of the European Spatial Development Perspective, which shows that though in the core of the EU most areas are within the influence of a major city or town, towards the peripheries there are wide zones that are only weakly connected to the European system of cities.

public transport links, including high-speed 'regional metros' such as those around Stockholm and Copenhagen, and planned for London, or even along true high-speed lines such as London–Ashford, Amsterdam–Antwerp or Berlin–Hannover. These would not of course be corridors of continuous urbanization, but rather clusters of urban developments, at intervals, around train stations and key motorway interchanges that offer exceptionally good accessibility. Some of these sites could be at considerable distances, up to 90 miles (150 kilometres), from the central metropolitan city. The UK's 2003 Sustainable Communities programme (Chapter 6) is an example of such a strategy.

In more remote rural regions, far from the global and sub-global centres, the pursuit of polycentricity must have yet another dimension: to build up the potential of both 'regional capitals' in the 200,000–500,000 population range (Bristol, Bordeaux, Hannover, Ravenna, Valencia), and smaller 'county towns' in the 50,000–200,000 range. The main agents will be enhanced accessibility both by road and (most importantly) high-speed train, coupled with investment in key higher-level service infrastructure (health, education); the systematic enhancement of environmental quality, to make as many as possible of these cities 'model sustainable cities'; and finally the competitive marketing of such cities as places for inward investment and relocation. Again, but on a smaller scale, the growth of such centres could be accompanied by a limited degree of deconcentration to even smaller rural towns within easy reach.

At whatever scale, spatial planning strategies cannot impose rigid blueprints. They can only suggest broad desirable directions; since the ESDP is advisory, and the principle of subsidiarity will apply, implementation will come mainly at national, regional and local levels. And there can be no firm guarantee as to outcomes: increasingly cities will compete directly in a global marketplace, and it can and should be no part of planning strategy to discourage this process. The EU will, however, play an increasingly valuable role in coordinating efforts at these other levels, and in managing a variety of funds which can help shape them.

The most urgent question at European level, perhaps, is how far and in what ways the new Spatial Development Strategy can influence the distribution of the structural funds. This would seem to require more fine-tuned geographical targeting than the approach embodied in the 2007–13 allocation, focused on particular types of areas which could most effectively serve the twin principles of a more polycentric Europe and of clustered collaborative development. The objective would be to enhance those qualities most likely to raise the competitive position of such centres: above all, accessibility to flows of people and information, and a high urban quality of life. The structural funds could play a crucial role here, as could other specific Community funds, not least the TENs (Trans European Networks), which are making important contributions to funding serious infrastructure funds.

Meanwhile, the principle of polycentric development has subtly morphed into another: *territorial cohesion*. Now embodied in the 2009 Lisbon Treaty, it earlier appeared in the Amsterdam Treaty when Michel Barnier was commissioner for regional policy. It is an essentially French concept, difficult to translate. At its simplest, it aims to promote spatial equity by maintaining key services, even in remote areas in order to achieve a harmonious allocation of economic activities. It has long historic roots – both within the European Commission, where it can be dated back to the Delors presidency of the 1980s, and in France, where it can be traced back deep to the mercantilist policies of Jean-Baptiste Colbert in the 1660s and 1670s. In fact, it is essentially the application on a European stage of concepts that formed basic building blocks of French policy even before the Revolution, but then became enshrined in the basic constitutional principles of the Republic. Central among these was the notion that the state has a basic responsibility for guaranteeing social solidarity. Thus capitalism in its pure *laissez-faire* form, as found in mid-nineteenth-century Britain, never took root in France: there, capitalist institutions were allowed to exist only within a planning and regulatory framework, ordained by a powerful central state. Under the Delors presidency, this notion was extended to the whole European territory, through a new stress on social cohesion. And, through the agency of DATAR, the French central planning agency, whose officials had a disproportionate influence within the Commission's regional division, DG-Regio, this logically acquired spatial dimension: the uniquely French concept of *Aménagement du Territoire*, again applied on a European scale.

This was an ideology shared by Franco–European politicians of different persuasions: Delors was and is a socialist, Michel Barnier, who in 2009 became the EU internal market commissioner, is a Gaullist. After 2000 it came into conflict with a new emphasis on economic competitiveness – the Lisbon agenda – associated with the presidencies of Romano Prodi and José Manuel Barroso. This conflict can be tortuously resolved: territorial cohesion can reduce market distortions, and thus help create a level competitive playing field for cities and regions. But essentially it represents a different philosophy, in which disinterested Euro-mandarins correct the vagaries of a globalized competitive world. In this interpretation, territorial cohesion represents a subtle and reluctant adjustment by DG-Regio to the new competitiveness agenda. Perhaps for this reason it remains disarmingly vague, expressed in the obscurities of French philosophical language rather than in hard realities of planning policy. And it is a long way from being able to generate guidelines for the distribution of ERDF. That remains the domain of hard politics: the ESDP was stripped of the maps that alone could have given it some operational force, so remaining a set of metaphors. And the same, at least for now, seems to be the fate of territorial cohesion. But that is perhaps the inevitable result of trying to develop coherent policies for one of the largest, and certainly the most diverse, unit of territorial governance in the world.

References and further reading

A useful geographical basis is provided by Hugh Clout, *Western Europe: Geographical Perspectives* (third edition, Longman, 1994). Good general introductions to the regional problems of Western Europe, with special reference to the EEC countries, are Hugh Clout (ed.), *Regional Development in Western Europe*, third edition (Wiley, 1986a); Hugh Clout, *Regional Variations in the European Community* (Cambridge University Press, 1986); Richard Williams, *European Union Spatial Policy and Planning* (Paul Chapman, 1996); Tony Champion, Jan Mønnesland and Christian Vandermotten, *The New Regional Map of Europe* (*Progress in Planning*, 46, pt.1) (Pergamon, 1996); David Shaw, Peter Roberts and James Walsh (eds.) *Regional Planning and Development in Europe* (Ashgate, 2000); Paul N. Balchin and Luděk Sýkora, with Gregory Bull, *Regional Policy and Planning in Europe* (Routledge, 1999) and Mark Tewdwr-Jones and Richard Williams, *The European Dimension of British Planning* (Spon, 2001).

On urban problems, Hugh Clout, *Europe's Cities in the Late Twentieth Century* (Royal Dutch Geographical Society, 1994); Vincent Nadin, Caroline Brown and Stefanie Duhr, *Sustainability, Development and Spatial Panning in Europe* (Routledge, 2000); Peter Newman and Andy Thornley, *Urban Planning in Europe: International Competition, National Systems and Planning Projects* (Routledge, 1996); and Gerd Albers, *Urban Planning in Western Europe Since 1945* (Spon, 2000). An excellent overview is Paul Cheshire and Dennis Hay, *Urban Problems in Western Europe: An Economic Analysis* (Unwin Hyman, 1989); it can be supplemented by the rather older books by Leo van den Berg, Rod Drewett and Leo H. Klaassen, *Urban Europe: A Study of Growth and Decline* (Pergamon, 1982) and Tony Fielding, *Counterurbanisation in Western Europe* (*Progress in Planning*, 17, pt.1) (Pergamon, 1982).

Three related volumes with a wealth of material on European spatial planning, all edited by Andreas Faludi and published by the Lincoln Institute of Land Policy, are *European Spatial Planning* (2002), *Territorial Cohesion and the European Model of Society* (2007) and *European Spatial Research and Planning* (2008).

8 Planning in the United States since 1945

To many Europeans, even well-informed ones, planning in the United States is a contradiction in terms. The country is seen as a land where rampant individualism provides the only guide to economic development or physical use of land. Planning, either in the sense of positive programmes for the regeneration of depressed regions, or in the sense of control over land use in the interest of the community, is thought to be virtually non-existent. Thus the United States is seen as a land where the phenomenally rapid settlement process has been accompanied by unprecedented destruction of irreplaceable natural resources; where extreme affluence marches hand in hand with large-scale pockets of poverty, often close by; where urban areas sprawl unregulated into fine open country, leaving a trail of ugliness and economic inefficiency. Fiercely critical as it may be, this is the stereotype which many European professional planners, and many intelligent European citizens, hold.

It contains both elements of truth and elements of complete distortion. Of course, pollution and destruction of resources and depressed regions and urban sprawl do exist – on a much larger absolute scale than in Western Europe. But at the same time the United States in the postwar era possesses a vast and complex system of planning agencies and of planning measures – both of a positive and a negative kind. Furthermore, just as in Europe, these operate at two distinct levels: first, the level of national/regional economic development planning; and second, the level of regional/local physical development planning. Both systems have had profound effects on the pattern of postwar economic and physical change in the United States; though, it can be said at the outset, some of them do not seem to have been very effective in relation to their scale and cost.

More perhaps than in any European country, because of the vast scale of the continental United States, these two levels of planning can be regarded as distinct. Indeed, one major recurring criticism of United States planning is that while economic planning tends to deal with very large regions, physical planning is exceedingly local and small-scale; the intermediate level of planning for the city region, though much written about and reported upon, is not very effective in practice. So the two scales of planning can usefully be treated at distinct levels.

Economic development problems

Though international comparisons are notoriously difficult and possibly misleading, it appears clear that regional disparities in economic development are somewhat greater in the United States than in typical Western European countries. In the United Kingdom, for instance, official statistics show that if the average national household income per head is set at 100, regional variations (in 2009) ranged from 125 (in the case of London) to 85 (in the North East) or 87 (in Northern Ireland). In the United States, the range of median household income in 2008 was from 166 (in Maryland, an essentially suburban state between Washington and Baltimore) to 73 (in the southern state of Mississippi). Of course, the size and geographical grain of the two analyses are very different; if the whole of Western Europe were taken as a more apt comparison with the United States, the discrepancies (as between Luxembourg or Hamburg at one extreme and the rural parts of Greece or Portugal on the other) would appear more extreme. But in addition it must be remembered that the American analysis is in terms of states, which are often large and very varied areas; there are great differences in economic development between the New York City area and up-state New York, between the Detroit–Flint axis and northern Michigan, or between the Dallas–Houston zone of Texas and the north-western part of that state. Whatever the scale and the grain, variations in levels of economic development and of personal income remain stubbornly large in the United States.

By and large, these variations can be related to the character of the economy. The high-income areas of the United States tend to be urban regions specializing in the newer, more technically sophisticated manufacturing or service industries; they include the major urban areas of the western states with their dependence on the aerospace industrial complex and on computing and control systems (Seattle, Los Angeles, Phoenix); the Silicon Valley area of central California; and the Texan cities of Houston, Dallas and Fort Worth with their combination of petro-chemicals and newer engineering industries, and advanced services. And these have been the areas in which population and employment have grown (Figure 8.1). Conversely, and more relevantly here, the low-income areas tend to be zones where the employment base is declining, either under the influence of falling demand of increasing efficiency of production; in some cases, too, the basic industry makes extensive use of rather low-skilled, poorly paid labour. One extensive group of such areas includes the agricultural regions of the south east (Old South) and south central parts of the country; these include both the former slave plantation areas which were converted to sharecropping of cotton (or to a lesser extent tobacco) after the Civil War, and the mountain areas of the Appalachians and Ozarks which have traditionally been inhabited by poor subsistence farmers. In many parts of this vast zone, which sweeps in a great crescent from the Carolinas through the Deep South states of Georgia, Alabama, Louisiana, Mississippi, Kentucky and Tennessee to Arkansas and eastern Oklahoma, average personal incomes are one quarter or less of the American national average; indeed, parts of these areas exist largely outside the mainstream of American life, resembling quite closely the traditional peasant economies of Europe. Overlapping geographically with this zone is another type of depressed area: the declining mining communities of the central Appalachians, extending from Pennsylvania and West Virginia down into eastern Kentucky. Within this zone, too, are many cities and towns dependent on iron, steel and heavy engineering, whose income is considerably lower than the national average. Other industrial areas where demand has fallen or competition from other regions has been severe may also exhibit the symptoms of decline; among them, the textile towns of New England are most notable. In the 1970s they were joined by other cities – most ominously, the car-manufacturing region around Detroit and the associated steel and engineering cities of the Midwest and north east.

This regional analysis, however, omits the important fact that income levels may vary locally. Characteristically, and increasingly, many American urban areas display the pattern of a low-income inner core – often extending over a fairly wide area, to embrace most, if not all, of the incorporated city, and consisting largely of those residential areas developed before the Second World War – surrounded by higher-income suburbs. The explanation of this pattern lies both in economic and in social causes. Though in most

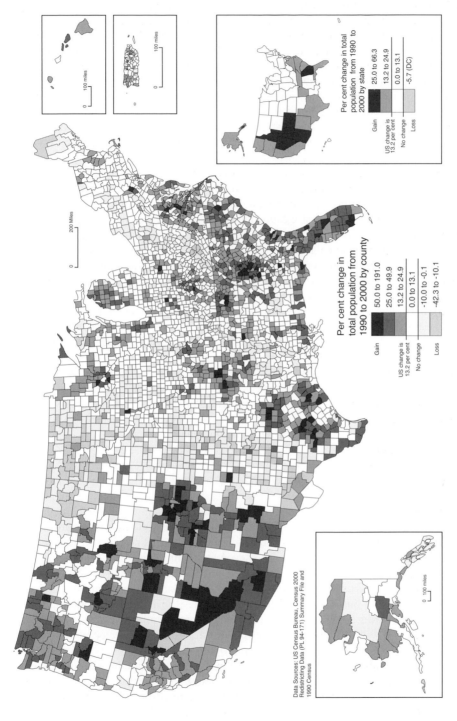

Data Sources: US Census Bureau, Census 2000
Redistricting Data (PL 94-171) Summary File and
1990 Census

Per cent change in
total population from
1990 to 2000 by county

Gain
50.0 to 191.0
25.0 to 49.9
13.2 to 24.9

US change is
13.2 per cent
0.0 to 13.1

No change
-10.0 to -0.1

Loss
-42.3 to -10.1

Per cent change in total
population from 1990 to
2000 by state

Gain
25.0 to 66.3
13.2 to 24.9

US change is
13.2 per cent
0.0 to 13.1

No change

Loss
-5.7 (DC)

Figure 8.1 Population change in the USA, by county, 2000–8. The map reveals a remarkable contrast between widespread population decline in the nation's rural heartland, and the explosive growth of urban areas in the mountain and desert west. Ironically, after the recession of 2008 some of these boom areas experienced a drastic reversal of fortunes, with out-migration, housing foreclosures and widespread property abandonment.

American cities there is still a considerable concentration of highly paid managerial and professional jobs in the central business districts of the cities, increasingly these jobs are migrating to suburban business or research centres; the better-paid, more highly skilled factory jobs have already decentralized in large part. In any case, even if the jobs remain in the cities, the people who work in them live in far-flung suburbs, commuting over increasingly long distances each day. Only in the 1990s was there an observable trend for some urban pioneers to return to live in restored inner-city areas immediately around the downtown core.

This is not a new phenomenon; it goes back almost to the beginnings of rapid urban growth in the United States, around the mid-nineteenth century. But since the Second World War it has accelerated under two influences: the suburban building boom, which will be described later in this chapter, and the mass migration of low-income, low-skill workers – in the 1950s and 1960s, African-Americans from the rural South; in the 1970s and 1980s, Latinos from Puerto Rico, Mexico and other parts of Latin America – into the inner areas of the great northern cities and some southern and western ones (San Francisco, Los Angles, Houston). Again, the latter phenomenon is not new; traditionally, poor immigrants to America first established themselves in the inner city, moving out as they acquired income and knowledge. But because the vast majority of the new immigrants are ethnically or culturally distinct, they suffer from prejudice in their attempts to follow this traditional route outwards; and since the higher-paid jobs move out while the lower-paid, less-skilled jobs tend to remain, they tend increasingly to be trapped in a vicious circle of inner poverty and lack of opportunity.

In summary, therefore, the contemporary United States presents a picture of islands of relative poverty amidst general affluence. Though it needs to be stressed that this is relative poverty – by world standards, the poor in America are certainly quite well off – nevertheless, for people actually experiencing it, this poverty will be intensely felt simply because people will judge themselves by the standards they see in their own society generally. These islands of poverty occur within two scales: first, on the broad regional level, in wide tracts of the southern and Appalachian States; and second, on the local level, within the inner cores of many metropolitan areas. These are the problem areas with which any development programme will need to deal.

Machinery for economic development

Until the 1960s the machinery for regional economic development in the United States tended to be quite local and *ad hoc* in character, even if it resulted from initiatives from the federal government. Thus in the 1930s and 1940s the great scheme for regenerating the Tennessee Valley, which runs through seven states in the heart of the South, was a highly successful piece of integrated development planning, which combined power, water supply, flood and erosion regulation, recreational planning, new industry and agricultural development; but, though a centrepiece of the Roosevelt administration's New Deal policy of the mid-1930s, it was not emulated on any similar scale elsewhere. After the Second World War, as in Britain, the impetus for regional development tended to fade under the influence of general full employment and widespread affluence. But again, as in Britain, there was a general realization that – at any rate in times of economic recession – many areas were not sharing in the general prosperity. Thus, by the early 1960s, two thirds of the labour-market areas of the United States had unemployment rates of 6 per cent or more.

The early attempts to deal with these disparities took a familiar form. The Area Redevelopment Administration (ARA), set up in 1961 within the Department of

Commerce to help areas with high unemployment or low local incomes, was empowered to grant loans to small businesses and to make loans or grants (up to 100 per cent of cost) in respect of public infrastructure; as a condition of help, a local region must prepare an overall economic development programme (OEDP). The Public Works Acceleration Act in 1962 gave the ARA still further funds for grants to provide infrastructure. But there was no overall strategy for distributing help from the centre, partly because there was too little research into needs and, because of constant political pressure, the constant tendency was for help to be spread too widely to achieve the necessary impact. There was an over-emphasis on expensive construction projects and not enough attack on the problem of aiding individuals to readapt themselves to economic change through retraining programmes.

By 1965 this criticism was being faced; the Public Works and Economic Development Act, in that year, converted the Area Development Administration into an Economic Development Administration (EDA), with bigger funds which were to be concentrated in a restricted number of major regions whose character was defined in the Act. First there were redevelopment areas with serious economic problems and with median family income 40 per cent or more below the national average; second, there were economic development areas, combining at least two redevelopment areas, and a substantial city capable of acting as a growth centre for them; and third, regional planning commissions could be set up for areas which overran state lines, such as the Ozarks, the Upper Great Lakes and the Four Corners (Arizona–New Mexico–Colorado–Utah) in the western mountains and deserts. Even then, though the experience was still that help was scattered among too many small areas and small cities; that most of the actual help was too concentrated on loan capital subsidies to businesses rather than direct help to the disadvantaged; that, where such help was available, it was geared too much towards training for the sorts of low-skill, low-pay jobs that were traditional in those depressed regions; that programmes often did not help the most disadvantaged groups; and that the machinery for implementation, which required widespread local agreement, was weak.

The Appalachian Regional Development Act of 1965, which established the Appalachian Regional Commission, in many ways represented a more hopeful experiment. Here, in this great upland mass which stretches from New England to the Deep South and which separates the huge concentration of population on North America's East Coast from the almost equally dense grouping in the Midwestern states, the problems were both physical and human. The natural resources of the region, once extremely rich – resources of timber, soil, coal and other minerals – had been removed by ruthless exploitation. The people were depressed and impoverished by centuries of isolation; their incomes were among the lowest in the United States and their standards of education, health and housing were often abysmally low. To grapple with these problems the commission supervised the injection into the region of no less than $679 million in the years 1965–9 (inclusive) alone. But it is significant that no less than $470 million went on an ambitious programme of super-highway construction, the avowed objective of which was to increase both the accessibility of the region to the outside world and the contacts within the different parts of the region itself. Doubtless this was a useful objective, but the concentration on expensive construction contracts seems to represent a real distortion of investment; once again, as so often in the history of American regional programmes, the benefit passed mainly to outside business interests rather than to the hard-pressed people of the region. It is only fair to say, though, that educational programmes (especially vocational training), health and housing improvements were also part of the total package; and that welfare of the population was a more prominent objective than in the ARA or EDA programmes earlier discussed. Administratively, too, the Appalachian programme broke new ground in developing joint federal–state developments; and it provided a model for other regional exercises in other depressed areas, such as the Ozarks, the Four Corners region in the mountainous areas of the West (where poverty was a problem among the many Indians on the reservations) and the coastal plains of the South where poor black sharecroppers had similar problems.

One important point about all these regional programmes – severely throttled back by the Nixon administration in 1973 – is that by earmarking aid for particular areas, they represented a conscious attempt to break away from the bad American tradition of spreading help thinly among all states and all areas, however different their problems. But all of them, like the ARA/EDA programmes, seem to have been conceived very largely in terms of the broader regional problem. By definition, in the United States this was difficult to solve because of the great size of such regions and their isolation from the areas of real economic dynamism; thus, though the Appalachian Commission made a conscious attempt to develop a policy of growth centres within its region, there were still many hundreds of miles distant from the main industrial areas of the country to which migrants were still moving. Almost ignored, on the other hand, was the growing and perhaps more readily soluble problem of localized poverty within the major urban areas. The great non-white migration from the rural South to the northern cities, which doubled the African American population in the northern states between 1940 and 1970, ironically provided a potential solution to the biggest of the major regional problems by transferring poor agriculturalists *en masse* to areas of greater economic opportunity; but only at the expense of distributing it in smaller pockets within the cities. Thus the nature of the problem shifted: in the 1960s it was still seen as a national/regional one, by the 1970s it was increasingly a central city/suburban one.

Metropolitan growth and change

The postwar United States, in fact, has witnessed three great human migrations: one, of African-American workers and their families from the mainly rural areas of the South to the northern cities; a second, of white families from the central cities of metropolitan areas to their suburban rings; and, in the last quarter of the twentieth century, a major wave of immigration from overseas, especially of people of Hispanic origin from Puerto Rico, from Mexico, from other parts of Latin America and from Cuba, from Latin America and across the Pacific, from East Asian countries like China, Taiwan, Korea, Hong Kong, Vietnam and the Philippines.

In quantitative terms, the second was the most important. Whereas 1,457,000 African-Americans migrated from the South between 1950 and 1960, and another 1,216,000 between 1960 and 1970, the corresponding estimates for migration to the suburbs – most of it white – were 5.8 and 4.9 million. Together, the three migrations resulted in a progressive occupation of northern cities by ethnic minority populations. In the year 2000 Detroit had an 82 per cent African-American population, Baltimore 64 per cent, Washington 60 per cent, Philadelphia 43 per cent and Chicago 37 per cent. More than 1.7 million Hispanics lived in Los Angeles, 46 per cent of the total population; more than 2.1 million in New York, where they constituted 27 per cent of the total. There was modest suburbanization of ethnic minorities in the 1980s and 1990s, but – save in a few contiguous older suburbs – seldom did the proportion exceed 10 per cent of the total.

The motive for both migrations was the same: improved economic and social status. But it operated in very different ways in each case. Whereas many of the African-Americans who migrated north did so out of necessity – their traditional economic base, sharecropping, had been suddenly removed by the development of cotton-picking machinery in the early 1950s – the new white suburbanites were voluntary movers in search of better housing and general environment. Furthermore, though the blacks moved unaided, the suburban white migration was powerfully assisted by federal policies. The Federal Housing Administration (FHA) had been created in 1934, a product of the Roosevelt New Deal; in 1949 the Housing Act established a Housing and Home Finance Agency (HHFA) to coordinate the activities of FHA and other official agencies. From the start, the emphasis of HHFA was on purchase of new homes; loans were easily available on a 10 per cent down payment basis from FHA, and interest rates were low at first. Further, FHA established standards of construction and of appraisal, which became current throughout the building industry and which improved the quality and reliability of new home construction.

Logically, the new housing was built on land that was previously undeveloped. The widespread use of the septic tank – a device which is normally restricted to rural areas in Britain – together with almost universal car ownership allowed a great deal of freedom in location; in particular, it meant that housing areas did not need to be as compact as in the interwar years. Thus sprawl developed in two ways: first, the house itself, and even more so its garden space (in American English, 'yard space') tended progressively to occupy more land, so that typical net residential densities dropped from ten to six and finally to between one and four houses to the acre; and secondly, the individual housing subdivisions tended to leapfrog, leaving areas of undeveloped land between them (Plate 8.1). Such far-flung development would have been inconceivable

Plate 8.1 Levittown-Fairless Hills, New Jersey, USA: postwar suburban development in the Atlantic urban region. Low-density single-family homes occupy subdivisions, with much leapfrogging of urban development over patches of vacant land. Commuting and movement generally in such areas depend almost exclusively on the private car.

without mass dependence on the private car; but in turn the pattern encouraged further scatter, since the new suburban areas were typically too far from the city to make use of its shops or services. Thus big new shopping centres developed in, or between, the new suburbs, rivalling the older urban centres in scale and generally excelling them in design (Plate 8.2). Jobs tended to decentralize too; in the 1950s and 1960s, blue-collar manufacturing jobs, associated warehousing and those service jobs that were tied to the needs of the suburban population; in the 1970s and 1980s, routine office work and even headquarters offices, drawn from their traditional downtown locales by the lure of cheaper land and the increasing availability of the labour force in the next-door suburbs. After the mid-1950s an ambitious programme of interstate highway (motorway) construction greatly eased the journeys of suburb-to-suburb commuters, further aiding the trend (Plate 8.3).

The suburban housing boom certainly performed a valuable service for many millions of Americans – in particular, those marrying and founding families, who made up record totals in the 1950s. Such people – ranging from highly paid managerial and professional groups through the range of white-collar clerical workers to the more-skilled factory workers – enjoyed solid benefits from life in suburbia, whatever popular sociology might say by way of criticism: good, well-built houses in pleasant neighbourhoods with congenial neighbours, good schools and convenient services. And the highly dispersed decision-making structure that guided the whole process did avoid massive social errors;

Plate 8.2 Milford Center, Milford, Connecticut, USA. Located in the fast-growing suburban zone outside New York City, this is a good example of the suburban, edge-of-town, new shopping centres that have developed on a large scale for car based shoppers in the United States since the Second World War.

like democracy in policies, it may have avoided the spectacular success, but it equally avoided spectacular failure. Finally, with constantly rising prices, suburban house buyers found their purchase was a useful hedge against inflation, especially when prices rose sharply, as in the mid-1970s.

The criticisms, in fact, are rather different. They are that the whole process could have been carried through so as to have given an equally good environment (or perhaps a better one) with less use of land and with lower resulting costs for public services, if the intervening undeveloped areas of land had been developed first; that sometimes the new suburbia was not as attractive visually as it might have been; and that, most seriously, the benefits have been denied to a substantial proportion of the total population. Comparing

Plate 8.3 Freeway interchange in Los Angeles. The southern California metropolis, with a population of over 6 million by 2002, has developed almost entirely in the era of mass car ownership since 1920. Thus it has grown quite differently from older cities, with wide dispersion of jobs and homes and generally a low density of development. Long-distance commuting is made possible by hundreds of miles of freeways which criss-cross the vast urban area, now 100 miles (160 kilometres) across.

average mortgage payments with figures of annual earnings, it is not difficult to calculate that the possibility of buying a new house has been beyond the capacity of at least the whole lower half of the income scale. True, many of these could still hope to buy second-hand houses in the older residential neighbourhoods of the central city or the inner suburbs. But a substantial proportion were condemned to live in rented housing which, because of failings in the tax laws, tended to be left by its owners to decay.

Generally, at the end of the first decade of the twenty-first century, housing in the United States had never been so affordable. The National Association of Realtors (NAR) found that in the third quarter of 2009, the median family annual income of $60,415 was more than 50 per cent higher than the qualifying income needed to buy a median-priced, existing, single-family house ($177,900) with a 20 per cent down payment – the highest level of housing affordability since the NAR started reporting in 1971. Some observers regard these figures as systematically over-optimistic. In any case, they do not distinguish between one metropolitan area and another. In the third quarter of 2009, the first-time buyer index stood at 105.3 per cent; in other words, the median-income family had just above the income needed to buy a median-price house. But in California, the index was only 64 per cent, and it varied from 85 per cent in the High Desert, deep in the interior of Southern California, down to 47 per cent in San Luis Obispo County and 49 per cent in the San Francisco Bay, one of the most expensive housing markets in the United States. In such areas, people are forced to search ever farther from their place of work, where prices are lower; and this lengthens the daily commuter trip over freeways that had once been free-flowing, but now were increasingly gridlocked. And to this must be added the problems of water supply and waste management and air quality, the loss of open space and rural qualities in huge swathes of land around the major metropolitan areas. The almost inevitable result was the growth of special-interest groups devoted to maintaining and enhancing the quality of environment, but also to stopping further development: the arrival of the NIMBY movement as a dominant political philosophy of the late twentieth century. Everywhere from New Hampshire and Virginia to the San Francisco Bay and the Central Valley of California, these problems of growth and spread came to dominate the life of the average American. But the result was to add one more turn to the screw, making it perversely even harder for new arrivals to enter the housing market.

In contradistinction to Britain and many other European countries, the United States does not cushion the lower-income group to any extent by providing new public housing; over the period 1945–70, less than 3 per cent of all non-farm housing starts were in the public sector, as against about 57 per cent in Britain. Nor, until the Housing and Urban Development Act of 1970, was there any federal funding to develop new towns on the British model. And, after the 1974 Housing and Community Development Act, the emphasis was not on replacing substandard housing – which had ceased to be a major problem – but on subsidizing low-income tenants who could not afford market rents: households whose rents exceeded 30 per cent of their incomes received aid, either on a project-by-project basis or – more radically – to live in privately rented housing, subject to a cap of 'Fair Market Rent' which varies according to local housing market conditions. Since 1999, this Housing Choice Voucher Program has been the basic means to finance low-income tenants – parallelling the trend in the United Kingdom and many other European countries, where voucher-type schemes have also become the primary means of subsidy.

By concentrating so heavily on house construction for sale in an inflating market, and by failing to provide a stock of new well-designed housing for lower-income groups, American postwar housing policies have in effect condemned a substantial minority to live in poor, run-down, overcrowded neighbourhoods (Plate 8.4). At the same time, successive attempts to upgrade these inner-urban areas did not have conspicuous success, with the exception of a number of major central business district redevelopments and a whole series of gentrified neighbourhoods, both of which ironically have displaced even more low-income and minority residents. And this could become even more of a trend, because – partly driven by the new migration from overseas – America's cities again recorded population increases in the 1990s, after many decades of decline. And this continued into the new century: between 2000 and 2008, more than 80 of America's top 100 cities grew: New York City by 4.3 per cent, Los Angeles by 3.5 per cent, Washington also by 3.5 per cent. Many sunbelt cities recorded substantial gains: Houston grew by 13.3 per cent, Phoenix by 18.2 per cent, San Antonio by 16.1 per cent and Atlanta by a remarkable 27.9 per cent. But interestingly, three major cities seriously impacted by deindustrialization, and with large minority populations, declined: Chicago recorded a loss of 1.5 per cent, Philadelphia 4.4 per cent, Detroit 3.5 per cent. These losses were dwarfed by flood-struck New Orleans, which lost over one third of its population, 35.5 per cent. But they demonstrated a deep and continuing pattern of shift from the rustbelt cities of the North East and Midwest, and the booming sunbelt cities of the South and West.

As a result, American society is becoming increasingly stratified by income, occupation and race. Even if the suburban development process cannot bear the whole blame for this, it must bear a part. In fact, many past federal programmes actually rebounded against the disadvantaged low-income inner-city resident; urban renewal programmes, carried through under the 1949 Housing Act, became synonymous in many cases with bulldozing the homes of low-income residents, and there was all too little provision of alternative housing for those displaced. The proposals for rehabilitation of existing housing under the 1954 Act – designed to meet criticisms of the earlier urban renewal

Plate 8.4 A ghetto area. This is fairly typical of the racial ghettos that exist on a large scale in the inner areas of many American cities. Black people – many of whom moved from the rural south after 1945 – found it difficult to escape into the suburbs, where the better housing and job opportunities have been located.

programmes – failed to have the expected impact on the condition of inner-city housing. In the 1960s, it is true, policies were redesigned to focus help on central city residents; more federal mortgage aid was concentrated on cheaper central city housing, and the federal government took the lead in trying to coordinate welfare and social service programmes for low-income families there.

This trend towards social planning, which is well marked in the 1970s, really indicates recognition that the problems of American low-income city residents – above all the non-white ones – have to be viewed as a whole; housing and physical planning form only a small part of the bundle of policies needed to deal with a complex problem. The great success story of the American economy in the 1980s and 1990s was that unemployment rates for all races fell by 40 per cent over the quarter-century between 1983 and 2008, from 9.6 to 5.8 per cent (with small upward blips in 1992, 2003 and 2008), and that the decline was particularly spectacular for African-American workers: from 19.5 per cent in 1983 to only 10.1 per cent in 2008 (with a low of 7.6 per cent in 2001), while the rate for whites fell from 8.4 to 5.2 per cent and that for Hispanics from 13.7 to 7.6 per cent. And interestingly, by the early twenty-first century unemployment rates were low even in cities that had been particularly impacted by deindustrialization: against a national rate of 6.5 per cent in October 2008, the Detroit metropolitan area recorded 9.4 per cent, Philadelphia 5.7 per cent, Cleveland 6.1 per cent and Chicago 6.2 per cent.

However, these overall figures concealed the critical importance of education to employment prospects in the knowledge economy. Between 1992 and 2008 the unemployment rate for those without a high school diploma fell only fractionally, from 11.5 to 9.0 per cent. Further, for such unqualified workers the only employment on offer tended to be minimum wage jobs with few or no career prospects. In 2008, 13.2 per cent of all households had incomes below the poverty line – the biggest-ever one-year jump recorded, further wiping out the gains made in the 1970s and 1980s, and a figure higher than in 1968. The child poverty rate, 19.0 per cent, represented 14.1 million children living in poverty; 35.3 per cent of all people in poverty were children. Worse, 24.7 per cent of blacks and 37.2 per cent of female-headed families were poor. From 2000 to 2008 black children in poverty rose from 31.1 to 33.9 per cent. For families of single mothers, the level rose from 33.0 to 37.2 per cent. Of the 8.1 million families living in poverty, 3.6 million of them were headed by single mums.

The result was remaining huge geographical concentrations of urban poverty. The poor were being increasingly segregated in islands of urban poverty, shut off from the mainstream economy and mainstream society (Plate 8.4). Many black families were thus caught in a vicious circle of poor job opportunities, poor education and family breakdown. It is small wonder that even though indices of social malaise – such as crime (especially violent crime), illegitimate births, drug abuse and poor health – have tended to fall in recent years, they are much higher in those areas where non-whites are concentrated.

To cope with such problems, the central cities found themselves facing a progressively larger tax burden. In particular, police and fire services, aid to dependent children and educational expenditures were disproportionately high in cities like New York, Chicago, Boston and Detroit. The share of local and state spending that is funded by the federal government nearly doubled from 1950 to 1975. But this mainly reflected highway expenditure; only recently has local welfare expenditure benefited much. The flight of richer people and of industry from the cities has left them with rapidly increasing needs and a shrinking tax base, and Washington has had to step in. Between 1957 and 1978 it is estimated that for a group of larger cities direct federal aid rose from an average 1 per cent of spending to a staggering 47.5 per cent. Many non-whites found themselves trapped in a vicious circle of social problems and rising expenditures from which they could not escape. Racial disturbances in the cities during 1967–8 intensified the desire of many blacks to leave, but hardened the barriers against them in the white suburbs. Continued migration from the South, coupled with a high rate of natural increase, made many major cities more than half black by 1980 – and this has persisted down to 2000, as has already been seen. To make matters even more problematic, during the 1970s and 1980s employment as well as white population was leaving the cities for the suburbs; not only did this intensify the cities' financial crisis, but it reduced the pool of well-paying jobs available to the black city populations within easy travelling distance. The 1990s saw a welcome upturn in centre city service jobs, but many of these are beyond the reach of low-skill workers.

To these dilemmas, two kinds of solution appeared in the 1970s and 1980s. The first was public–private partnerships, in which cities allied with private developers – with major injections of federal and state money in the form of grants for public works, subsidies linked to private leverage and tax exemptions including the designation of enterprise zones, as well as new institutional forms such as development corporations – to regenerate a major part of their decayed inner-city area, often an old port area or an abandoned railroad freight yard, via a major construction-plus-rehabilitation project. The classic models were Baltimore's Inner Harbor (Plate 8.5), Boston's Quincy Market and Waterfront, San Diego's Horton Plaza and a score of imitators. (In Britain, the same phenomenon was observable on an even bigger scale in the London Docklands.) Critics might argue that this was simply urban renewal all over again – indeed, in Boston and in Baltimore one follows the other in an unbroken line – but now the ambitions are greater: against a background of unprecedented deindustrialization and urban decline, the transformation of decayed industrial and port cities into leading centres of the new service economy, through a combination of producer services, theme-park entertainment, leisure shopping and street theatre. It can be criticized, indeed has been criticized, but it may be the only effective way of bringing employment to the most deprived inner-city ghetto areas. The same could be said for an innovation of the 1990s: Business

Plate 8.5 Baltimore Inner Harbor. One of the outstanding cases of urban revitalization in the core of an old American industrial and port city, this festival marketplace is now one of the biggest tourist attractions in the United States.

Improvement Districts (BIDs), pioneered by Philadelphia but now widely imitated in other American and British cities, whereby businesses paid special contributions to improve the physical condition of downtown areas in order to compete with the managed spaces in suburban retail malls. In many cities these have extended to adjacent residential areas which have been restored as gentrified enclaves, contributing to the population increases observed by cities in the 1990s. But there is an important caveat: in many cities, the downtowns are now islands surrounded by decayed or decaying areas. True, here too there are encouraging signs of rehabilitation and repopulation, partly fed by immigrants from abroad. But the picture is everywhere still partial and even contradictory.

This raises a final and disturbing question. As the new information technologies permit ever more distant decentralization of urban activities, what is the role of the traditional central city? Even if some cities manage to survive through their special qualities – New York and Chicago and Los Angeles as major world centres, Boston and San Francisco as centres of education and technology, culture and tourism, Atlanta and Dallas and Denver as regional nodes – can all of America's older cities survive? Or do they represent some historic anomaly, destined to disappear like the ghost towns of the American West? Or is their fate to become theme-park museums which recall the places they once were? (See Plates 8.6, 8.7 and 8.8.)

Plate 8.6 Urban regeneration in California: Mission Bay. One of the largest urban regeneration schemes in the United States, this is the old dockland area of San Francisco, now almost completely lost to the container port of Oakland on the other side of San Francisco Bay. The mixed-use scheme includes a major extension of the San Francisco Muni Metro light rail system, a new baseball park and the campus of the University of California, San Francisco (UCSF), one of the leading centres of medical research in the United States.

This relates to the problem of governance. The United States has had as little success as most other countries in remodelling its local government structures to grapple with the metropolitan problems which face it. The local pressures against change have been too strong; and, in the nature of the American system, the leverage exerted at the centre has been too weak.

One solution to the problem of the cities was somehow to bring the cities and the suburbs into some sort of closer relationship. As in most other countries, the political geography of the twentieth-century United States has long ceased to represent social or economic reality. City boundaries have been hardly extended for half a century, during which time suburban expansion has extended the effective urban area many times. From the start, it suited many suburban communities to go their own way and make their own rules; in the 1950s and 1960s, as the cities plunged into their vicious circle of poverty and civic bankruptcy, to maintain independence became for the suburbs a matter of survival. Consequently, though intellectual voices were raised in favour of metropolitan governments which would plan city and suburbs as a single unit for the common good of both, real-life experiments in this direction were few. Only Greater Miami went for full-scale metropolitan government, while Minneapolis and its suburbs adopted a looser form of federation. In other areas, like San Francisco, regional government initiatives

Plate 8.7 Millennium Park Chicago. Opened in 2004 on former railway yard and parking lots, the Park has become the civic centre for downtown Chicago. Costing $475 million, the park contains several public buildings and works of art including the Frank Gehry designed Jay Pritzker Pavilion, and the Cloud Gate installation by artist Anish Kapoor. Widely applauded for its successful urban planning and design, the park has also attracted criticism for the way it controls tightly the use and commercialization of the space.

Plate 8.8 Kentlands, Maryland: a New Urbanism project outside Washington, DC, designed by Andres Duany and Elizabeth Plater-Zyberk, one of the architectural-planning practices which have led the movement back towards traditionally designed higher-density residential development. Impressive in its urban quality and economic use of land, like other such developments it suffers from lack of integration with urban public transport: it is located 3 miles from the nearest DC Transit station and there are currently no plans for an extension.

foundered during the 1970s in the face of opposition from suburban localities – though San Francisco was making another effort at the start of the 1990s. Elsewhere, voluntary organizations like the Regional Plan Association in New York tried to produce strategies to guide regional growth that would depend on the agreement of a whole host of public and private agencies – a daunting task (Figure 8.2).

Planning powers and planning policies

This raises the critical question of the machinery of planning, its geographical basis and its effectiveness. To discuss this for a European readership is difficult, because in many ways the American system of government is unlike that in other countries. In the first place, it is federal; and traditionally matters of domestic importance, which would certainly include planning and local government, have been matters left to the states to determine. (The township system of government in the New England states, for instance, is quite different from the county system used elsewhere.) The power and influence of the federal government in domestic affairs, especially through the use of federal funds, has admittedly increased very strikingly since the Second World War – in this particular area of interest, above all in the 1960s. But state differences must constantly be borne in mind. Secondly, American local government is typically less tidy and more complex than European; services are supplied by a multiplicity of *ad hoc*, single-purpose agencies, such as planning commissions, boards of education or sewer commissions, so that a citizen may live within the area of a score of different local government units, some of them with different boundaries. Since these agencies are separately controlled (and separately elected) there is no logical reason for them to cooperate; very often they are at loggerheads. Coupled with the very strong role of private agencies in the urban development process, this means that there are very many more different agents or actors associated with urban growth and change than in the typical European situation; a fact that makes the whole process both more difficult to describe, and more difficult in practice to control.

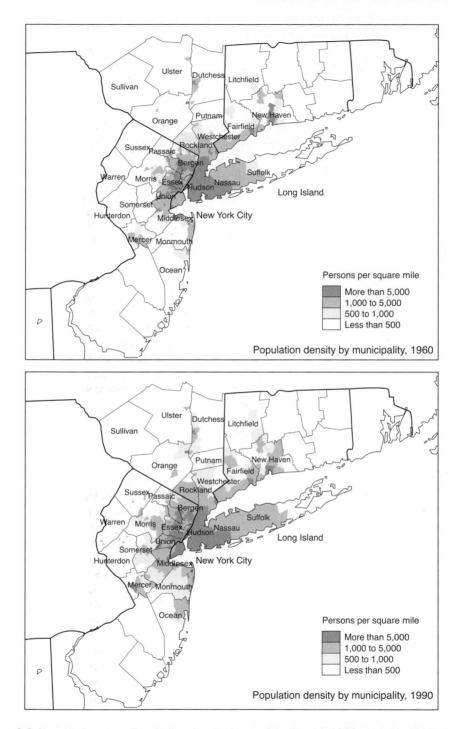

Figure 8.2 New York region. Population density by municipality: (a) 1960; and (b) 1990. Over 30 years suburbanization has extended far outside the core cities of the region. The Atlantic seaboard from Boston to Washington – 'Megalopolis', in the term of the geographer Jean Gottman – is one of the greatest massings of humanity in the world, with over 30 million people. It is an example of a highly complex multinuclear urban structure, where the growth of one city impinges on that of another.

At the top level there is the federal government agency for housing and planning: the Department of Housing and Urban Development (HUD). This is the nearest equivalent to a European department such as Britain's old Ministry of Housing and Local Government before its incorporation in the Department of the Environment; it deals neither with transportation, which is handled by a separate Department of Transportation (DOT), nor with national parks, which since their inception have been traditionally a responsibility of the Department of the Interior. But perhaps the most significant point about HUD is that it was set up only in 1966, after fairly bitter political opposition; up to that time, there was actually no central agency at federal level handling the complex of problems presented by urban growth. HUD inherited the responsibility for a variety of agencies concerned with housing, themselves brought under the umbrella of the Housing and Home Finance Organization in 1949; but perhaps more importantly, it was given extensive new responsibilities in the field of metropolitan area planning, which it has pursued with energy. In the years after its creation, HUD first became responsible for the Model Cities programme authorized by Congress in 1966, under which cities were aided by Washington to adopt a comprehensive, across-the-board, integrated approach to problems of housing, renewal, job training, education, health and welfare in poor city neighbourhoods. It linked spending on physical renewal and environmental rehabilitation to the provision of social services and housing, and it aimed to increase the capacity of local governments to take action, but it was hindered by implementation problems and spending cuts. Then in 1968 HUD turned to the creation of new communities with federal aid. The Housing and Urban Development Act of that year made funds available to cover the difficult transitional period when heavy investments were needed but when returns were low. A further Housing and Urban Development Act in 1970 gave modest extra funds for new community creation. The specific aims were to channel a significant part of future metropolitan population growth – estimated in 1968 as 75 million by the year 2000 – away from contiguous suburbs and towards reasonably self-contained communities; and to work towards a social balance by ensuring reasonable proportions of low-income and non-white residents. A particular part of this programme would create new towns in-town through public–private partnership; a bold experiment, in practice it achieved disappointing results.

The most significant development of the early 1970s, under the Nixon administration, was revenue-sharing. It aimed to shift implementation of programmes to state and local governments, thus eliminating cumbrous federal bureaucracy. But its immediate effect was hard for the bigger, declining cities: their funding increased, but less rapidly than that in the southern and western cities and less rapidly than their needs. However, two other important new programmes of this time – the Comprehensive Employment and Training Act (1973) and the Community Development Block Grant (1974) – later proved important for the cities. The first gave them help in physical reconstruction, including commercial and industrial facilities; the second provided for job creation and training. By the late 1970s, they were getting 40–50 per cent of the total funds for economic development, physical development and fiscal support. But the Reagan budget cuts in 1981 threatened to slash these programmes and to exacerbate the plight of the declining cities at a time when their economic base was badly eroded.

HUD can exert considerable leverage on local governments across the United States by its control over federal funds. Even before it came into existence, it had been laid down that to obtain federal highway moneys, local areas would have to engage in a comprehensive planning exercise. Similar federal funds were available for wider metropolitan planning under Section VII of the 1954 Housing Act; later, progressively during the early 1960s, comprehensive metropolitan plans were made mandatory for any authority that required federal funds for a wide variety of purposes – whether for sewers, open space, education or urban renewal. But HUD does not have the same power as its British counterpart: the power to require local authorities to submit plans (and regular revisions of those plans) to it for approval. Nor may it designate land for a new community, with consequent restrictions on the amount of public liability for compensation, as happens in Britain. As so often in American government, the powers are permissive rather than regulatory.

Indeed, it would be difficult to see how this could be otherwise. Even national programmes funded largely by federal funds, such as the 41,000-mile (74,000-kilometre), $41,000 million Interstate Highways Program, must be executed by the individual state governments. Powers of compulsory purchase for the creation of new communities would have to be exercised with the states' approval, through their courts. Local government structure itself, as previously indicated, varies from state to state. Any federal department is, therefore, necessarily more circumscribed than its British equivalent.

At local government level, two complications obtrude. The first is the multiplicity of agencies; this means that even where a number of separate boards or commissions operate over the same geographical area (usually a county) their operations are not likely to be coordinated in any way by a central managing unit, as would occur in the average British or European local government unit. Thus the sewer commission may have as big a potential influence on urban development as the so-called planning commission; so may the commissioner for highways; but all these are separate agencies, each going its own way independent of (and sometimes in spite of) the others. The other complication, exceedingly difficult to grasp for the average European, is that the use of land may be affected by two different operations, planning and zoning; but that these two are in principle (and not seldom also in practice) separate. In 1968 there were over 10,000 local government units in the United States with a planning board or similar organization; but the great majority of these had either no staff, or a completely inadequate one, and the plans they prepare generally lack legal status or binding power. If there is a central governing board for the country, that board will not be governed by the decisions of the planning board; nor, of course, will any specialized agency. In this situation the sewer agency may be in effect the real planning authority rather than the nominal planning board.

In fact the real core of the American system of land-use control is not planning, but zoning. But it is formally separated from the planning system; it is administered by a separate zoning commission for each local authority area, it need take no account of the plan (if any) and it is essentially a limited and negative system of control over changes in land use. By definition, zoning is a device for segregating different types of land use, usually on a rather coarse-grained basis. The traditional view has always been that zoning cannot in practice altogether stop a potential developer; s/he must be left with some profitable development of the land. This, essentially, is because the American system – in contradistinction to the British one – does not involve any method of compensation for lost development rights, such as was embodied in Britain's historic 1947 Town and Country Planning Act. Rather, the American zoning system rests on the concept of police power – a term hardly known in Britain – which is a general residual power of government to pass laws in the interests of general public health, safety and welfare. Zoning, in a fairly rough-and-ready way, has achieved some of the same objectives in practice as land-use planning in Britain; it has segregated land uses thought to be incompatible, such as factory industry and homes. But by definition it could not easily protect open countryside against development; that could usually be assured only by public purchase as a national or state park or similar facility, a practice which was used widely around certain metropolitan areas, especially San Francisco, to create de facto green belts by the 1970s. In actuality, zoning is more subject to abuse than land-use control in Britain; notoriously, if the landowner or prospective developer is persistent enough, s/he can usually get the change s/he wants. So, with ineffective planning and only semi-effective zoning, controls over the physical growth and change within urban areas are much weaker in the United States than in Britain – or many other parts of Europe. Developers and, behind them, the consumer of their products are still sovereign in a way that in Britain they are not.

However, during the 1970s some striking changes occurred in this regard. As already noticed, certain cities, especially in California, attempted to enact growth control ordinances; and, after considerable legal wrangling, some of these were declared constitutional. The main tool is control over the servicing of land for development, which is then used to reach an agreement with builders for an annual growth of new homes, with an agreed proportion in the moderately priced category. Elsewhere, communities have zoned agricultural land as very large building lots – 40 acres (16 hectares) and more – which precludes suburbanization. The State of California's Williamson Act gave farmers a reduced property tax (rating) burden provided they covenant to protect their land from development for ten years. By the 1980s these measures had been joined by fiscal devices: impact fees, whereby developers are compelled to pay local communities for the indirect costs imposed by development, such as local roads and schools; and development agreements, similar to the British planning agreements, whereby they make such provision directly.

New urbanism and smart growth

Old-style renewal programmes continue to operate, but in a radically new way: the HOPE VI Program, introduced in 1992 and recognized in federal law in 1998, specifically provides funds to replace low-grade housing by new mixed-income developments. As of 2005, the programme had distributed $5.8 billion through 446 federal block grants to cities, with the highest individual grant being $50 million. However, reflecting the new emphasis on transferring aid from projects to people, much of this funding has gone to provide vouchers for lower-income residents to live in the reconstructed neighbourhoods. The novel element in HOPE VI is its stress on replacing monolithic high-rise apartment blocks, characteristic of so many public housing schemes of the 1950s and 1960s, with lower-rise traditional housing with a strong emphasis on single-family terrace houses facing directly onto the street. This reflects the pioneering work of the social anthropologist Oscar Lewis, who invented the concept of defensible and indefensible urban space, and his influence on the new urbanism movement.

In the 1970s Lewis had researched the design of housing, especially public housing, from the viewpoint of crime and personal safety. He found that many public housing schemes actually invited crime, both against the person and against property, because of design faults – in particular, a lack of the basic oversight by neighbours, which had been characteristic of traditional neighbourhood streets. This resonated with the campaign led by Jane Jacobs, a journalist who had campaigned against urban renewal in the Greenwich Village area of her native New York City, and whose hugely influential 1961 book, *The Death and Life of Great American Cities*, had likewise called for retaining traditional street patterns. Following their lead, in the late 1980s a small group of architect–designers – notably the partnership of Andres Duany and Elizabeth Plater-Zyberk, on the east coast, and Peter Calthorpe in California – began to design new neighbourhoods on very traditional lines that had been characteristic before the Second World War, with fairly dense terrace houses facing out onto grid-pattern streets, enjoying good-quality public transport. This, they say, is human-scale and walkable, with varied land uses and good public spaces. They stress ordinances to reintroduce traditional kinds of neighbourhoods – examples of which can be found in California, Virginia and Florida.

New urbanism is certainly radical in rejecting the sprawling four-houses-per-acre, totally automobile-dependent model of the suburbs that spread around American cities in the 1950s and 1960s, though less so in California where – contrary to the myth – suburbs around Los Angeles and San Francisco were built much more compactly because of topographic constraints. It is also surprisingly close in design terms, especially density, to earlier models of urban design like Radburn in New Jersey or Greenbelt in Maryland, in the late 1920s and early 1930s. The main difference lay in the new urbanists' insistence on traditional street grids and their consequent emphasis on 'permeability' for traffic as well as pedestrians – a radical rejection of the previous idea that people on foot should be protected from direct exposure to cars.

But these say nothing about the regional relationships of the neo-traditional developments to the wider metropolitan area. Peter Calthorpe's West Coast version tackles this by grouping development around a combined commercial-transit core: Transit-Oriented Developments (TOD) or what two other planners, Michael Bernick and Robert Cervero, have called transit villages. These are developments deliberately designed at densities higher than conventional automobile-oriented suburbia, that have shopping and other essential daily services within easy walking distance, and that are above all grouped around good-quality transit. But, with a few exceptions (such as Calthorpe's work in San Jose), most actual examples of new urbanism are either new suburbs on greenfield land (Laguna West in California, Kentlands in Maryland) or resort/retirement communities (Seaside and Celebration in Florida). And their commendable stress on good public transport was too often not achieved in actual practice. One of the most celebrated examples by Duany and Plater-Zyberk, Kentlands in the Maryland suburbs of Washington, DC, is located 5 miles beyond the end of the Metro line – and there are no plans for an extension. Calthorpe's plan for a new suburb on the site of the old Stapleton airport in Denver, Colorado, did feature an extension of the city's light rail system – but this is now not certain.

To handle this problem, a far more integrated approach was necessary, combining land use and transport planning, and with powers to control the kind of low-density suburban sprawl that has characterized so many areas over the last half century. Unsurprisingly, smart growth became the great American planning mantra of the 1990s and 2000s. Starting in a small way as early as the 1970s, with a few statewide planning and growth management programmes, by 1999 it had embraced no less than 100 laws in 27 states. But the practical effects so far have been minimal: between 1981 and 1999 America built 27.3 million new homes, especially in the sunbelt and West; and in the 39 largest metropolitan areas, 80–85 per cent of this construction has been suburban, though recently there has been much more interest in older cities. Thus, the so-called new urbanists argue, most of this development is very unsmart growth: it is happening in suburbs where it is only possible to move by car because low densities make public transport non-viable, lacking multi-family housing and without any recognizable centre.

Smart growth advocates therefore say that there must also be negative growth controls, in particular an urban growth boundary: a line defining the edge of the metropolis, based on land capacity to house a growing population. This technique was first used in the 1970s in Portland, Oregon. But experts are not agreed: the new urbanists Andres Duany and Elizabeth Plater-Zyberk argue that it has proved over-generous in permitting sprawl, while others assert that it has brought price rises – a charge denied by Peter Calthorpe. An alternative approach is the Countryside Preserve, which sets aside conservation land, preferably as continuous green belts; classic examples are the regional park system in the San Francisco Bay Area, brought under public ownership mainly because it was vital watershed land, and Boulder in Colorado, which from 1967 taxed itself to buy a green belt that now covers an area twice that of the city itself but that has created a housing shortage, forcing lower-income people out and increasing daily commutes. Yet a third approach, now adopted in Florida, Maryland and New Jersey, is the Urban Service Boundary (USB), outside which the state will not finance infrastructure extensions such as water or sewerage. The critical question for all these options is whether individuals will be willing to take the steps necessary to achieve smart growth, be it a high-density housing project, less parking or tax increases. One observer has argued that this will require stronger central planning, higher taxes and denser development than Americans have traditionally accepted: a critical difference as compared with densely populated Western Europe.

Some conclusions

So some key figures are arguing for a new direction in American urban growth, and it must be said that local voters seem to be backing their ideas: many people now berate the waste of land and the inefficiency represented by continuing low-density, leapfrogging urban sprawl, the profligate use of natural resources, and the pollution and cost in lives, resulting from the widespread dependence on the private car; the lack of choice, and the homogeneity of standards, brought about by suburbanization: the ugliness which may result from failure to control the more bizarre manifestations of commercialism; and the decay of the larger, older cities in the northern and eastern parts of the nation.

Whether this will result in different kinds of urban growth will depend on their capacity to accept some of the consequences: as one observer has commented, it is easier to talk the talk than to walk the walk. And, while conceding the force of all these arguments, it is worthwhile to recapitulate the arguments in favour of process of mass suburbanization: the fact that it has provided so many millions of ordinary middle-level-income people with good housing in pleasant neighbourhoods with good services, all at a price within their means. Indeed, as we have seen, the main argument against it is not that it is bad in itself, but that it could have been better done, and that its benefits have been denied to a large – and probably increasing – proportion of the entire population. Those shut out from the suburbs – meaning the poor and the minorities, which are too often synonymous – would be the first to argue in favour of life in the suburbs, if only they could get it.

The other conclusion is that at the wider regional level, the vast cost of the various development programmes has not yielded anything like a satisfactory return. Far too much funding has been spread indiscriminately across the country, both among areas in great need and areas in less need. This is because a philosophy of economic development, based on careful analysis of goals and objectives, has not been clearly worked out at the centre. Such a presentation of objectives would need to take into account the often conflicting considerations of geographical relationships, management of natural resources and the conservation of the environment. From it would emerge – hopefully – a set of guidelines as to the regions and areas where growth should be positively encouraged, those where no particular aid was needed, and even those where growth should be positively discouraged for various reasons – whether of conservation, congestion or simply lack of economic prospects. Only against this background could the federal government begin to pursue a policy of selective aid through support of educational, health and job training programmes. Up to now, this has not been done in a clear or conscious way.

As a result of these failures, there is no doubt that the contemporary United States – perhaps to a greater extent than any Western European country – presents strange anomalies which must be regarded as failures of urban policy. On the one hand, widespread diffusion of a remarkably high level of material wealth; on the other, minorities living in poverty which is striking just because it is so far below the general level. On the one hand, massive construction achievements in areas such as suburban housing and new highways; on the other, paralysis and decay in the inner cities. On the one hand, general private affluence at a level not witnessed elsewhere in the world; on the other, in places, real public squalor in the form of blighted landscapes and urban decay. These are contrasts of which increasing numbers of Americans are aware, but the remedy is still hard to find. Millions are voting with their feet: central cities have shrunk, entire major metropolitan areas have stagnated, and the sunbelt of the South and West is gaining massively at the expense of the frostbelt – or, in other words, the traditional industrial belt (the North East) and the Midwest. There are now signs of a reversal: the upturn in central city populations, the success of some examples of the new urbanism in the commercial market. But they are small counter-signs to set against the big long-term trends of dispersal and suburbanization. Urban America is a land in transition, and at the start of a new millennium no one, not even the experts, is sure where it is going.

Further reading

An excellent recent summary of demographic and social trends is Reynolds Farley, *The New American Reality: Who We Are, How We Got Here, Where We Are Going* (Russell Sage Foundation, 1996).

The standard textbook on American planning is J. Barry Cullingworth, *Planning in the USA: Policies, Issues and Processes* (Routledge, 1997). It can usefully be supplemented by two books, coincidentally by British civil servants: John Delafons, *Land Use Controls in the United States* (second edition, MIT Press, 1969) and Richard Wakeford, *American Development Control: Parallels and Paradoxes from an English Perspective* (HMSO, 1990).

For regional economic planning, the standard source is Lloyd Rodwin (see Further reading, Chapter 7); this also deals in passing with the urban problem. More detail on economic planning is provided by John H. Cumberland, *Regional Development Experiences and Prospects in the United States of America* (Mouton, 1971). A wealth of material on the background to American regional development is found in Harvey S. Perloff, Edgar S. Dunn, Eric E. Lampard and Richard F. Muth, *Regions, Resources and Economic Growth* (Johns Hopkins University Press, 1960; paperback edition available).

Marion Clawson, *Suburban Land Conversion in the United States: An Economic and Governmental Process* (Johns Hopkins University Press, 1971), remains the standard source on postwar American urban growth: Marion Clawson and Peter Hall, *Planning and Urban Growth: An Anglo-American Comparison* (Johns Hopkins University Press, 1973), draws on this source to provide an account for the general reader of contrasts between the British and American patterns of postwar urbanization.

On the changes in the regional and urban geography of the United States in the late twentieth century, key sources are George Sternlieb and James Hughes, *America's New Market Geography: Nation, Region and Metropolis* (Center for Urban Policy Research, 1988); Ann Markusen, *Regions: The Economics and Politics of Territory* (Rowman & Littlefield, 1987); Lloyd Rodwin and Hidehiko Sazanami (eds.) *Deindustrialization and Regional Economic Transformation: The Experience of the United States* (Unwin Hyman,

1988); and Carl Abbott, *Urban America in the Modern Age: 1920 to the Present* (Harlan Davidson, 1987). Anthony Downs, *New Visions for Metropolitan America* (Brookings Institution, 1994), is an excellent study by one of America's leading experts in public policy.

Jane Jacobs' *The Death and Life of Great American Cities* (Random House, 1961; Jonathan Cape, 1962) became an instant planning classic, still well worth reading today. On the new urbanism, see Peter Katz, *The New Urbanism: Toward an Architecture of Community* (McGraw-Hill, 1994); Andres Duany and Elizabeth Plater-Zyberk, *Suburban Nation: The Rise of Sprawl and the Decline of the American Dream* (North Point Press, 2000); Peter Calthorpe and William Fulton, *The Regional City: Planning for the End of Sprawl* (Island Press, 2001) and John A. Dutton, *New American Urbanism: Re-forming the Suburban Metropolis* (Skira, 2000).

Christopher Jencks and Paul E. Peterson (eds.), *The Urban Underclass* (Brookings Institution, 1991), provide an excellent overview of the economic and social problems of America's cities. It can be supplemented by William J. Wilson, *When Work Disappears: The World of the New Urban Poor* (Knopf, 1996). As an antidote, see Paul Grogan and Tony Proscio, *Comeback Cities: A Blueprint for Urban Neighborhood Revival* (Westview Press, 2000), which analyses the recent demographic turnaround in American cities and some of the reasons for it.

9 The planning process

Up to now this book has been an introduction to the problems and the content of spatial planning, treated historically. That has been its aim, as the preface indicated. But now, this last chapter tries to make a bridge to the actual process of planning, as it is carried out by progressive planning authorities at the present time. This process is based strongly on theoretical concepts, which are well set out in modern textbooks of planning. Therefore, this chapter, which tries to distil the central content of these more advanced texts, will perhaps serve as an introduction to them for the student of planning.

We need in this to distinguish three quite separate stages in the evolution of planning theory. The first, developed from the earliest times down to the mid-1960s – and well exemplified in the early development plans coming after the 1947 Town and Country Planning Act – could be called the master plan or blueprint era. The second was ushered in from about l960, and replaced the first approach through the Planning Advisory Group (PAG) of 1965 and the 1968 Town and Country Planning Act; it could be called the systems view of planning. The third, which began to evolve in the late 1960s and the 1970s, is more heterogeneous and more diffuse; it may best be labelled the idea of planning as continuous participation in conflict. In what follows, we shall first describe the transition from blueprint to systems planning, and second, the more complex transition to participative–conflict planning.

Systems planning versus master planning

The change that occurred after 1960 was based on the notion that all sorts of planning constitute a distinct type of human activity, concerned with controlling particular systems. Thus spatial planning (or, as it is called here, urban and regional planning) is just a subclass of a general activity called planning; it is concerned with managing and controlling a particular system, the urban and regional system. It follows from this that all planning is a continuous process, which works by seeking to devise appropriate ways of controlling the system concerned, and then by monitoring the effects to see how far the controls have been effective or how far they need subsequent modification. This view of planning is quite different from the one held by an older generation of planners, such as Geddes or Abercrombie, or even the generation which set up the planning system in Britain after the Second World War. These older planners saw planning as concerned with the production of plans, which gave a detailed picture of some desired future end-state to be achieved in a certain number of years. It is true that under the 1947 Planning Act in Britain, deliberate provision was made for review of the plans every five years. But the philosophy behind the process was heavily oriented towards the concept of the fixed master plan.

Arising from this basic difference of approach, there were also detailed differences between 1940s and 1960s planning. The old planning was concerned to set out the desired future end state in detail, in terms of land-use patterns on the ground; the new approach, embodied in Britain in the new structure plans prepared under the 1968 Planning Act, concentrates instead on the objectives of the plan and on alternative ways of reaching them, all set out in writing rather than in detailed maps. Again, the old planning tended to proceed through a simple sequence, derived from Patrick Geddes: survey–analysis–plan. The existing situation would be surveyed; analysis of the survey would show the remedial actions that needed to be taken; the fixed plan would embody these actions. But in the new planning, the emphasis is on tracing the possible consequences of alternative policies, only then evaluating them against the objectives in order to choose a preferred course of action; and, it should be emphasized, this process will continually be repeated as the monitoring process throws up divergences between the planner's intentions and the actual state of the system.

The new concept of planning derived from one of the newest sciences: cybernetics, which was first identified and named in 1948 by the great American mathematician and thinker Norbert Wiener. Rather than dealing with a completely new subject matter, cybernetics is essentially a new way of organizing existing knowledge about a very wide range of phenomena. Its central notion is that many such phenomena – whether they are social, economic, biological or physical in character – can usefully be viewed as complex interacting systems. The behaviour of atomic particles, a jet aeroplane, a nation's economy – all can be viewed, and described, in terms of systems; their different parts can be separated, and the interactions between them can be analysed. Then, by introducing appropriate control mechanisms, the behaviour of the system can be altered in specific ways, to achieve certain objectives on the part of the controller. The point here is that it is necessary to understand the operation of the system as a whole (though not necessarily in complete detail throughout) in order to control it effectively; unless this is done, actions taken to control one part of the system may have completely unexpected effects elsewhere. A good example is the design of a motorcar; if the designer produces extra power without considering the total impact on the rest of the complex system that makes up the car, the result could be instability or rapid wear of other parts, with disastrous results.

Cybernetics has already had considerable practical applications in modern technology, especially in the complex control systems which monitor spacecraft or automatic power stations. Its applications to the world of social and economic life are still tentative. Some observers think that human mass behaviour is too complex and too unpredictable to be reduced to cybernetic laws. Others find ethically repellent the idea that planners should seek to control the operation of machines. All that can be said with certainty is that in some areas where people and machines interact – as for instance in urban traffic control systems – cybernetics is already proving its effectiveness. It still remains to be proved definitively whether the application can be extended equally well to all areas of human behaviour.

Fundamental to the concept of systems planning – as the cybernetics-based planning has come to be called – is the idea of interaction between two parallel systems: the planning or controlling system itself, and the system (or systems) which it seeks to control. This notion of constant interaction should be kept in mind throughout the following account of the systematic planning process. More particularly, we are concerned with this process as it applies to spatial planning, using the word 'spatial' in its widest sense: it need not be limited to the three-dimensional space of Euclidean geometry, but may extend for instance to include notions of economic space (the costs involved in traversing distance), and psychological or perception space. Nevertheless, there can be little doubt that in some sense, however distorted by psychological or economic factors, the relationship of parts of the urban and regional system in geographical space must be the central concern of the urban and regional planner.

To control these relationships, in a mixed economy such as the United States or the countries of Western Europe, the planner has two main levers: one is the power to control public investment, especially in elements of infrastructure such as roads, railways, airports, schools, hospitals and public housing schemes; the other is the power to encourage or discourage initiatives from the private sector for physical development, through incentives or disincentives to industrial development, controls on land use and environmental regulations. Both these forms of power, of course, vary in their scope and effectiveness from one nation or society to another. Different countries invest different proportions of their gross national product in public infrastructure (though in advanced industrial countries there are limits to this variation); different nations have very widely differing controls over physical development (though in none, apparently, is there either a complete lack of such controls, or a completely effective central control). Therefore, almost by definition, the urban or regional planner will never be completely ineffective, or completely omnipotent. The planner will exist in a state of continuous interaction with the system s/he is planning, a system which changes partly, but not entirely, due to processes beyond the planner's mechanisms of control.

Against this background, it is now possible to appreciate the schematic summaries of the planning process set out by three leading British exponents of the systematic planning approach: Brian McLoughlin, George Chadwick and Alan Wilson. McLoughlin's account (Figure 9.1a) is the simplest; it proceeds in a straight line through a sequence of processes, which are then constantly reiterated through a return loop. Having taken a basic decision to adopt planning and to set up a particular system, planners then formulate broad goals and identify more detailed objectives which logically follow from these goals. They then try to follow the consequences of possible courses of action which they might take, with the aid of models which simplify the operation of the system. Then the planners evaluate the alternatives in relation to their objectives and the resources available. Finally, they take action (through public investment or controls on private investment, as already described) to implement the preferred alternative. After an interval they review the state of the system to see how far it is departing from the assumed course, and on the basis of this review they begin to go through the process gain.

Chadwick's account of the process is essentially a more complex account of the same sequence (Figure 9.1b). Here, a clear distinction is made between the observation of the system under control (the right-hand side of the diagram) and the planner's actions in devising and testing his control measures (the left-hand side). Appropriately, there are return loops on both sides of the diagram, indicating again that the whole process is cyclical. But at each stage of the process, in addition, the planners have to interrelate their observations of the system with the development of the control measures they intend to apply to it.

Wilson's account (Figure 9.1c) is even more theoretically complex, but again it can be related to Chadwick's. In it, there are not two sides of the process which interact, but three levels presented vertically. The most basic level, corresponding to part of Chadwick's right-hand sequence, is simply called 'understanding' (or, in the terminology of the American planner Britton Harris, 'prediction'). It is concerned wholly with devising the working tools, in the form of techniques and models, which are needed for the analysis of the system under control. The intermediate level, corresponding to another

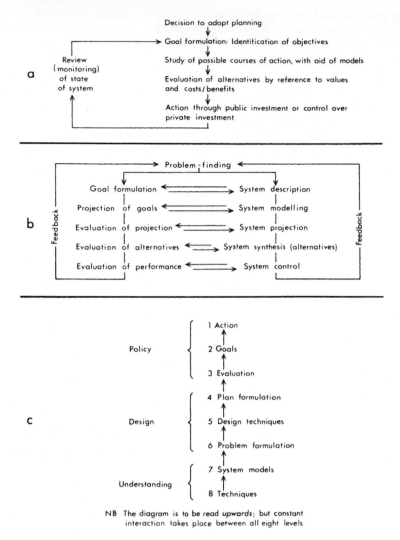

Figure 9.1 Three concepts of the planning process: (a) Brian McLoughlin; (b) George Chadwick; (c) Alan Wilson. During the 1960s interest developed in systematizing the process of planning, with a new stress on modelling and evaluating alternative designs or courses of action. These formulations drew heavily from the sciences of cybernetics and systems analysis.

part of Chadwick's right-hand side, is concerned with the further use of these techniques in analysing problems and synthesizing alternatives which will be internally consistent. The upper level, corresponding roughly to the left-hand side of the Chadwick diagram, is essentially concerned with the positive actions which the planner takes to regulate or control the system: goal formulation, evaluation of alternatives and actual implementation of the preferred alternative.

All three accounts are helpful ways of looking at the planning process. But since simplicity must be the essence of this summary chapter, the following accounts of the separate stages of the process are based principally on the classification of Brian McLoughlin.

Goals, objectives and targets

Planning, as a general activity, may have one objective or many. There is no necessary relationship between the scale and expense of a planning programme, and the complexity of the objectives behind it; thus the American moon-shot programme, one of the costliest pieces of investment in the history of humankind, had a fairly obvious single main objective. Most urban and regional planning activities, however, have multiple objectives. The first step in the planning process, then, is to identify these purposes which the planner seeks to achieve, to order them in terms of their importance and to consider how far they are reconcilable each with the other. This might seem obvious, yet surprisingly most plans of the past prove to be very perfunctory in their treatment of objectives; it seems almost as if the aims of the plan were so well understood that no one needed to set them down. But unless objectives are made explicit, no one can be sure that they are shared by the people they are being planned for; nor is it possible rationally to prefer one plan to another.

Modern plan methodology, therefore, lays great stress on this first step in the process. In particular, it distinguishes rather carefully between three stages in the development of aims: goal formulation, identification of objectives and target setting. Goals are essentially general and highly abstract; they tend to fall into broad categories such as social, economic and aesthetic (some of which categories may overlap), and they may include qualities of the planning process itself, such as flexibility. Some authors, notably Wilson, define goals in a rather different way, as areas of concern: in this view, the planner starts by identifying broad functional sub-systems which are of interest to him, because they appear to present problems which may be amenable to the controls he proposes to manage. Examples of areas of concern would include public health, education, income and its distribution, mobility (both physical and social) and environmental quality. Objectives in contrast are rather more specific; they are defined in terms of actual programmes capable of being carried into action, though they fall short of detailed quantification. They also require the expenditure of resources (using that word in its widest sense, to include not merely conventional economic resources but also elements like information) so that they imply an element of competition for scarce resources. Thus if 'mobility' is a general goal, the resulting objectives might include: a reduction of travel time in the journey to work, an improvement in the quality of public transport (or of a part of it) or a programme of motorway construction to keep pace with rising car ownership. Notably, as in the cases just quoted, objectives can only be devised as the result of a more detailed scanning of the system being planned, in order to identify specific malfunctioning or deficiencies. Finally, as a further stage of refinement, objectives are turned into targets representing specific programmes in which criteria of performance are set against target dates. Thus the detailed targets developed from the above objectives might include: construction of a new underground railway line within ten years to reduce journey times in the north-western sector of the city by an average of 20 per cent; or construction of a new motorway link within five years in order to cut traffic delays by some specific amount. Targets, by their nature, tend to be very specific and particular; one problem that emerges from the whole goals–objectives–targets process, therefore, is that of integrating rather disparate individual programmes into a coherent plan.

Already, this first stage in the planning process involves great difficulties of a conceptual and technical nature. In the first place, it is not entirely clear who should take the lead in the process. Broad goals for society, it might be argued, are a matter for the politicians, though the professional planner can play a valuable role by trying to order the choices. But politicians are largely involved with acute short-term issues; their timescale is very different from that of the planner, whose decisions may have an impact for generations. The public themselves form a very heterogeneous mass of different groups, whose value systems are almost certainly very different if they are not in open contradiction. Even the identification of these groups poses difficulties, because most people will belong to more than one group for different purposes; they will have interests and values as members of families living at home, as workers in a factory or office, as consumers, and perhaps as members of voluntary organizations, and the values of these groups may actually come into conflict with each other. Public opinion polls and other surveys may throw limited and distorted light on preferences, because most people find difficulty in thinking about highly abstract goals that do not concern them immediately, and because they will not easily imagine long-term possibilities outside their immediate range of experience. Because of differences and even conflicts of view, it is almost certainly impossible even to devise a satisfactory general welfare function, which would somehow combine all the individual preferences and weightings of different individuals or groups.

It is no wonder, then, that in his comments on goal formulation Chadwick points out that 'the gap between theory and possible practice is pretty wide'. Planners do the best they can by trying to amass as much information as possible about their clients and their values; by trying to identify acknowledged problem areas, where by fairly common agreement something needs to be done; and by using logical argument to proceed from general goals to more specific objectives. Evolving research tools, such as simulation and gaming – whereby members of the public are faced with imaginary choice situations which test their preferences – will also help to throw light on one particular dark area: the weighting of different objectives and the trade-off between them. But it should not be expected that there will be a dramatic breakthrough in this intellectually very difficult area.

Forecasting, modelling and plan design

Having defined objectives and given them some precise form in the shape of targets based on performance criteria, planners will turn to description and analysis of the urban or regional system they wish to control. Their aim here is to find ways of representing the behaviour of the system over time – both in the recent past, and in the future – in such a way that they can understand the impact of alternative courses of action that are open to them. To do this, the planners will produce a model of the system (or, more likely, a number of interconnected models which seek to describe the behaviour of its sub-systems). A model is simply a schematic but precise description of the system, which appears to fit its past behaviour and which can, therefore, be used, it is hoped, to predict the future. It may be very simple: a statement that population is growing by 2 per cent a year is in effect a model of population growth. But it may be, and often is, computationally quite complex.

There are two important questions that the planner needs to resolve about the modelling process: first, what aspects of the urban system one wishes to model; second, what sorts of model are available. The answer to the first question will, of course, depend on the planner's precise interests; the planner must first say what questions the model is required to answer. But usually, the urban and regional planner is concerned with the spatial behaviour of the economy or of society. In particular, the planner is interested in the relationships between social and economic activities – such as working, living, shopping, and enjoying recreation – and the spaces (or structures) available to house them. The planner will need to know the size and location of both, as well as the interrelationships between activities (transportation and communication) which use special spaces called channel spaces (roads or railways, telephone wires). Together, these aspects of the urban system can be said to constitute activity systems. Particularly important among them, for the urban planner, is the relationship between workplaces, homes, shops and other services, and the transportation system that links these three.

The answer to the second question – the choice of type of model – will again depend on the object of the planning exercise. Models, whether simple or complex, are capable of being classified in a number of different ways. They may be deterministic in character, or probabilistic (i.e. incorporating an element of chance). They may be static in character, or dynamic. Many of the best-known urban development models are static; that is, they project the system only for one future point in time, at which point the system is regarded as somehow reaching equilibrium. This, of course, is a totally unrealistic assumption which is not supported by knowledge of how the system actually behaves, and one of the main challenges is to produce better dynamic models which are useable. Another separate but related question is whether the model chosen is to be simply descriptive of the present (or recent past) situation, or predictive of the future, or even prescriptive in the sense that it contains some element of built-in evaluation. Self-evaluating models are not very common in urban and regional planning, though they do exist: the linear programming model, which automatically maximizes the achievement of some variable subject to certain constraints, is the most notable example and has been used in planning contexts both in the United States and in Israel. But more commonly the model merely predicts the future; it can be run a number of times with different policy assumptions underlying it, but finally the choice will be made through a quite separate evaluative process.

Yet another question is the choice between spatially aggregated models and spatially disaggregated models. A model which projects some sub-system for the town or region as a whole is termed spatially aggregated; a model which examines the internal zone-by-zone allocation of that system is spatially disaggregated. Urban and regional planners, of course, require both sorts of model, but the results of their spatially disaggregated models must accord with the control totals given by the spatially aggregated ones. Well-known population projection models, such as the cohort survival model (which operates through the survivorship rates of successive five-year age cohorts of the population), are spatially aggregated; so are the common economic models, such as input–output models. Models which predict future distributions of people and service industries within urban areas, such as the well-known Garin–Lowry model used in many planning studies, are, of course, spatially disaggregated.

Some models also combine an aggregated with a disaggregated element; this is true of Garin–Lowry. This model (Figure 9.2) starts with an assumed amount, and an assumed distribution, of basic industry – that is, industry the produce of which is exported from the city or the region, and which thereby provides an economic base or support for the people of that region. The model then calculates simultaneously both the aggregate amount, and the spatial distribution, of residential population and of the local service industry employment which is dependent upon that population. The aggregate totals are obtained by using two simple ratios, a basic employment–population ratio and a population–service industry employment ratio. The distributions are obtained by using a so-called spatial interaction model which, like most of this type is derived from the well-known gravity theory. This states that the interaction between any two areas which form part of a wider set of areas is directly proportional to their sizes (as defined, for instance, in terms of employment or population concentrations) and inversely proportional to the distance between them. Such a model contains a number of parameters and constants, which are values capable of being altered so that the model provides the best possible fit to the observed past or present facts; this process of fitting is called calibration of the model.

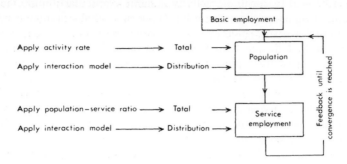

Figure 9.2 The Garin–Lowry model structure. First developed in the United States but employed extensively in Britain, Garin–Lowry is the best known of the mathematic models used to project the amount and distribution of residential population and local services. It depends on prior knowledge or assumptions about the amount of basic employment.

The Garin–Lowry model, then, starts with a simple distribution of basic industry – however this is defined for the purpose of the exercise. It ends with a picture of the urban area at some future point in time, showing the patterns of residential population, of service provision and of the work journeys and service journeys which link up these varied activities. It is capable of being run with different planning policy assumptions in it: different assumptions, for instance, about the distribution of basic industry, or of the pattern of transportation facilities which will affect the accessibility of the systems to each other. This, plus its relative simplicity and economy, have made it one of the most commonly used models in Britain and North American urban planning practice. Its chief disadvantage – that it is a simple, one-shot model requiring constant repetition to make it fit a dynamic planning framework – may be overcome within the next few years by the development of an operational dynamic version.

Model design is one of the most complex and intriguing stages of the modern planning process. Designing a model, or models, to suit the precise problem involves logical analysis of a set of interrelated questions. Once it is determined precisely which questions the model is supposed to answer, the problem is to list the concepts to be represented, which must be measurable. It is also necessary to investigate which variables can be controlled by the planner, at least in part; if the assumption is that no parts are controllable, then the model is a pure forecasting model, but if at least some of the factors are under the planner's control then this is a planning model. The planner must also consider what behavioural theories about systems are to be embodied in the model. The planner must consider technical questions, such as how the variables shall be categorized or sub-divided (as, for instance, population can be categorized by age, sex, occupation or industry group); how explicitly time will be treated; and how the model is to be calibrated and tested. The answers to these questions will depend in part on the techniques that are available, and on the relevant data that can be used to illustrate them, as well as on the computational capacity of the computer which will be used to run the models. Fortunately, with the increasing power of personal computers, this tends to be no longer a constraint.

Plan design and plan evaluation

Many standard accounts of the modern planning process refer to a stage which is called plan design, or plan formulation. To the layperson, this would appear to be the critical point where, when all technical aids have been used to the utmost, the planner takes command and exercises his/her creative abilities, just as s/he did in a simpler age before computer modelling had become an integral part of the planning process. In an important way, this is true: there must be at least one point in the whole process, and in all probability more than one point, where the planner exercises a power to synthesize disparate elements into a coherent plan. But in fact, this power has to be manipulated in close relationship to the machine. What the computer – and above all the personal computer – has done is to speed up, many times, the power to generate and to evaluate, alternative formulations of the plan. The capacity to design is essentially the capacity to use this power critically and creatively.

The design process, therefore, really starts as soon as the planner begins to design the models. At that point, the critical questions – what elements of the urban system should the models represent, and in how much detail – will finally determine the content of the plan design. To all intents and purposes, the model is the design, and alternative assumptions built into the model generate alternative design possibilities. Of course, the word 'design' here is not being used in a conventional sense. In most cases, the urban and regional planner does not end by producing a blueprint for actual physical structures on the ground. What the planner tries to do is to specify a future state or states of the urban and regional system which appear, from the operational model, to be internally coherent and consistent, and to be workable and feasible; and which also best satisfy the objectives which have been set. The content of the design, and of the model which embodies it, will depend on the focus and the objectives of the planner. Thus, if the plan stresses transportation, it will chiefly consist of a design for channel spaces to accommodate projected traffic flows. If the plan stresses social provision, it will embody locations for social service facilities in relation to the distribution of projected demands from different sections of the population. Invariably, following the modern stress on planning as a process, the design will not be a one-shot plan for some target date in the future; rather, the model or models which incorporate the design will represent a continuous trajectory from the present into the predictable future.

Design, therefore, essentially consists of two elements. The first is the choice of system models to represent the main elements which the design should incorporate, and the running of these models to give a number of coherent and realistic pictures of the future state of the system through time. The second is the process of evaluation of the alternatives to give a preferred or optimum solution. At the stage of evaluation, the goals and objectives which the planner has generated are applied directly to the alternative simulations of the future system.

Like most other terms in the planning process, the word 'evaluation' needs careful definition. To most lay observers, it conveys a connotation of economic criteria: evaluation, crudely, represents the best plan or money. Many notable modern planning exercises have in fact made extensive use of economic evaluation procedures; some of these will be described in summary a little later. But essentially, evaluation consists of any process which seeks to order preferences. Strictly speaking, it need not refer to money values, or to use of economic resources, at all.

What is essential is that evaluation derives clearly from the goals and objectives set early on in the planning process. The first question must be how well each design alternative meets these objectives, either in a general sense, or (preferably) in terms of satisfying quantified performance criteria. Very commonly, it is found that many objectives contain an element of contradiction in practice. It is difficult for instance to reconcile the objective 'preserve open countryside' with the objective 'give people the maximum freedom to enjoy the private environment they want', or alternatively to reconcile 'provide for free movement for the car-owning public' with 'preserve the urban fabric'. Somewhere along the line, either in the original formulation of objectives or in the evaluation process, it is necessary for the planning team to devise weights which rank some objectives above others, and indicate how much different objectives are worth in relation to each other. This may involve a conscious decision to favour one group of the client population more than another, because quite often the interests of these groups are in conflict: car owners versus non-car owners, for instance, or old-established rural residents versus new interests. Such value judgements are hard to make, and the political process must inevitably have a large hand in them.

To try to make plan evaluation more rigorous, since about 1955 at least three techniques have gained widespread currency in the planning world. The best known of these among the general public, cost–benefit analysis, is explicitly economic in its approach. It assumes that the best plan will be the one which delivers the greatest quantity of economic benefits in relation to economic cost; these latter being defined, as is usual in economic analysis, as alternative opportunities forego. (A simpler form of economic analysis, cost-effectiveness analysis, assumes that benefits from alternatives are equal, and analyses merely the variable cost; it is of limited use in urban and regional planning). Essentially, cost–benefit analysis is useful in situations where decision-makers want to know which of several alternatives represents the best economic value, but where normal market measures are not available. The business leader in private industry has no such problem: s/he can predict the demand for his or her product or service in the market, and so calculate expected return on capital invested. But public decision-makers have no market as a guide: they are producing services which are not sold at a price. Cost–benefit analysis, therefore, works by trying to create 'shadow prices' for items outside the market. The value of a road investment is defined in terms of savings in petrol, tyres, drivers' and passengers' time, and reductions in accidents; these last are valued in terms of lost capacity for earning wages, and on this basis even the value of a death in a road accident can be calculated in money terms.

This approach, however, throws up many problems – some so intractable that critics claim cost–benefit analysis to be of very limited use, and even positively harmful, in planning decisions. Valuing people's time, or the risk of accidents, in terms of wage rates may mean that poor people (and housewives, and children) are valued less than rich people, especially businessmen. Many important elements in planning, such as the value of a fine landscape or of an old building, are almost literally imponderables: there is no easy way that a value can be put on them. If an attempt is made to do so – landscapes can be valued in terms of the lengths of journey that people make in order to look at them, and old buildings can be valued in terms of insurance value put upon their possible destruction – many people will argue that it is ethically wrong to use such commercial judgements in such situations; the result of following the approach consistently, they say, would be that no building or landscape could ever be preserved if there were a good economic case for removing it, so that a motorway could be driven with impunity through Westminster Abbey, or a new airport for London could be located in Hyde Park. These very fundamental objections are closely related to another: cost–benefit calculations have to be applied to the planner's models of the future of the system, and if these models prove to be wrong in even small particulars, this may seriously affect the outcome of the analysis. In the celebrated controversy surrounding the Roskill Commission inquiry into the siting of London's third airport (1968–7), for instance, the cost–benefit analysis developed for the commission contained a very large element for the value of air travellers' time, and this in turn was highly sensitive to assumptions made about the future pattern of travel by air in Britain. Critics argued that it would be unwise to reach firm conclusions on such speculative (and easily upset) projections. Fundamentally, the objection to cost–benefit analysis is that it is too arbitrary in character. By trying to represent all types of costs and benefits, to all groups in the population, in terms of a single aggregate metric, it conceals the very considerable value judgements that underpin it behind an appearance of value-free objectivity. To some extent cost–benefit analysis can meet this criticism by producing sensitivity analyses; these show the impact of altering some of the basic assumptions in the analysis, and allow the decision-maker to consider just how much s/he would be willing to sacrifice of one element in order to achieve another. Cost–benefit analysis, in this argument, is not a magic touchstone but an educative device, which makes the decision-making process more rigorous by stressing the economic argument about the costs of alternative choices. But this does not meet completely the counter-argument about imponderables.

The second best-known evaluative device in planning, Nathaniel Lichfield's Planning Balance Sheet, specifically tries to deal with this criticism. It is essentially a modified cost–benefit analysis which tries to render in economic terms those items which are capable of being treated in this way, but which resorts to simpler devices for the imponderables. Unlike cost–benefit analysis in the strict sense, it makes no attempt to render all values in a common metric; it does not produce a 'rate of economic return', as cost–benefit analysis does, and it is not, therefore, very suitable for comparing a range of different investments. It is, however, specifically devised for the consideration of alternative plans for the same urban or regional system, and has been successfully applied to problems of urban renewal and of new town construction; a modified version of it was proposed by the British Advisory Committee on Trunk Road Assessment in 1977 for assessing motorway and trunk road plans, and is now used in all such cases. Later, Lichfield has developed it further into an even more comprehensive framework, Community Impact Analysis. Its merits are that it is highly disaggregative, stressing advantages and disadvantages of different plans for different groups in the population; and that it spells out its value-assumptions very carefully, so that the decision-maker is aided without having the decision taken out of his or her hands. Its disadvantage lies in its inevitable complexity, which means that the decision-maker needs a strong effort of will to question each successive weighting that is made in the course of the exercise; if s/he she fails to do this, the planner will tend to accept the weightings or trade-offs made by the professional evaluator, which the planner or the electorate may not necessarily share.

The Goals Achievement Matrix of Morris Hill, third of the evaluation devices which have gained currency in urban and regional planning, tries to deal with this problem by starting from the agreed objectives which the plan-making machine sets up. It compels decision-makers to make specific judgements about the weights they attach to the various objectives; these judgements are then applied to further judgements as to the degree to which alternative plans meet these objectives, expressed on a numerical scale. Like Lichfield's method, Hill's matrix recognizes that different groups of the public may have different value systems, so that they may place quite different weights on different objectives it allows for this by disaggregating its analysis. As with the Lichfield method, which it so closely resembles, the chief defect of the Goals Achievement Matrix is its complexity. But it has to be recognized that plan evaluation is bound to be a complex and controversial process

Most serious planning exercises now use some form of systematic plan evaluation technique, though they may not go as far as employing one of the three methods just described in its full rigour. Many are content with a considerably simplified version of the Goals Achievement Matrix, in which alternatives are judged against a check list of objectives, with a simple attempt at weighting. To this, more recently, has been a requirement for some form of Environmental Impact Analysis, at any rate for public projects such as new roads or airports. The development of policies and plans, meanwhile, are increasingly subject to a form of sustainability appraisal or strategic environmental assessment. Many in addition try to involve the public in the process of evaluation, by trying to obtain the view of a sample of the public on the question of the weights to be applied to different objectives, as well as on their preferences among the plan alternatives which have been generated. These pioneer attempts at public participation are open to the objection that many ordinary people cannot easily appreciate abstract qualities, such as flexibility or environmental quality, especially when they are applied to rather large-scale, diagrammatic plans which do not make specific reference to the local areas that people really know and understand. But they represent a beginning.

One important question about the whole plan-design process is whether it should be linear or cyclical. The version so far developed in this chapter is linear: that is, the alternative plans are developed and modelled, all in equal detail, up to the point where they are all evaluated side by side with a common set of evaluation procedures. In fact several major British planning exercises have instead used a cyclical approach. A number of very crude alternatives are developed, modelled and evaluated. Certain among them are eliminated, one or more are retained, and these (or combinations and permutations of them) are developed and modelled in greater detail. This process may be repeated three or four times, with the modelling-evaluation process progressively testing finer and more subtle variations of detail. The cyclical or recursive approach appears more complex, particularly when it is applied in the plan report. But it can be argued that it is more economical of the planning team's skills and of computer time, and by logically eliminating alternatives and concentrating on detailed variations it acts as a systematic educative process for the team.

Implementing the plan

By systematic evaluation of alternatives, the planner can select a preferred course of action for implementation. But it needs to be stressed again that this is no once-for-all decision. In the planning process outlined here, the whole exercise of modelling, evaluation and selection is continuously repeated. The objective is to have on the one hand a monitoring system which checks the response of the urban and regional system to the various planning measures which are taken to control its progress; and on the other hand the control system itself, which responds flexibly and sensitively to the information controlled by the monitoring system. The analogy, of course, is with piloting a ship or an airplane. A course is set; a battery of instruments confirms that the craft is on course, or that it is deviating from course; appropriate control devices, either automatic or manual, take appropriate corrective action. The monitoring system thus tests the correspondence (or lack of correspondence) between the real-world situation and the model (or 'navigation chart') that has been set up to describe it. If there is a divergence, then either controls must be operated to bring the real-world situation again in conformity with the model design, or the model must be altered to make it a more realistic description of the way the world works, or some combination of the two (Figure 9.3).

The above is frankly a description of a planning ideal rather than of present planning reality anywhere. The world that urban and regional planning seeks to control is much bigger and richer in content than the rather limited piece of reality represented by the course of a ship or an airplane. To reduce it to schematic terms by means of a model is correspondingly more difficult, and the likelihood of error much greater. Because of the complexity of the human resources involved, the control systems open to the planner are much cruder and less effective than those available to the ship's master or airline pilot. The history recounted earlier in this book proves definitely that even in strong

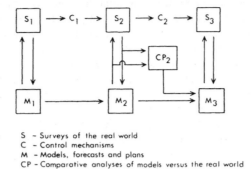

S – Surveys of the real world
C – Control mechanisms
M – Models, forecasts and plans
CP – Comparative analyses of models versus the real world

Figure 9.3 The plan implementation process, according to Brian McLoughlin. This formulation, in which models interact with surveys of the real world through comparative analyses, and in which control processes are then applied to the real-world situation, draws heavily on analogies and insights from the sciences of cybernetics.

and effective planning systems, the world changes in all kinds of ways that planners fail to predict, so that plans may fail grievously to correspond to reality, even after a very few years. In any event, even if we did know how to control the responses in the urban and regional system, to exert pressure effectively might prove politically unfeasible.

In practice, as revealed in some notable planning controversies of recent years, a tidy systems view of planning may go wrong in a variety of ways. In the first place, knowledge about the external environment of the planning decision may increase rapidly, with unpredictable results. The changing economies of nuclear and conventional power production may invalidate a programme of power-station location; variations in the noise emission levels of jet airplanes, and improvements in ground level city-to-airport transport, may completely change the framework of a decision about airport location (while a change in the size of planes may make a new airport unnecessary); the development of quieter cars or even road surfaces, might render some of the current controversies about new road building irrelevant. In practice it ought to be possible to predict technical changes and their impact rather better than is now generally done. (It seems extraordinary, for instance, that just after World War Two, when jet aircraft were already flying, their noise impact seems to have been ignored in planning all the major civil airports of the world). But even so, it must be recognized that there will always be a considerable element of unpredictability and chance.

Second, plans can go wrong because of the complex interrelationships between different levels of the planning system, and between different elements in the planning situation. Thus a general, high-level strategic policy may be laid down by a national or regional planning authority, for apparently good cause, but with unexpected effects at the more local planning level. In Britain, office development policy was a good example: it was introduced in 1964–5 with the aim of restricting office growth in London and other major cities, and of promoting decentralization to new towns and development areas. But the restrictions had the effect of holding up for many years some important pieces of redevelopment in London, such as Piccadilly Circus, which depended for their commercial viability on the office content. The process can, however, work in reverse. Thus, though almost everyone was agreed on the need for a national motorway to relieve traffic congestion on the old road, work on the new highway was held up for over ten years as one local amenity society after another successfully diverted the line of the road from its own area. The fact that, as finally built, the motorway probably follows the least environmentally damaging line is perhaps some consolation.

Third, there is the fact that over time human values – or at least the values of those actively concerned – tend to change. In recent years there is evidence that the place of such changes is actually increasing; fashions in planning tend to change almost as fast as fashions in clothing. Since complex plans inevitably take time to prepare and then to execute, the result may be controversy. Urban redevelopment provides a good example. In the late 1950s and early 1960s, the key word was 'comprehensive redevelopment': to provide a better environment and separate people's activities from the danger and pollution of traffic, it was necessary to make a clean sweep of many old urban areas. But by the late 1960s and early 1970s, there has been virtually a reversal: influenced by the highly persuasive book by Jane Jacobs, *The Death and Life of Great American Cities*, the key words became 'conservation' and 'urban spontaneity', and younger planners in particular wanted to keep the chaos and disorder of the older city, which they saw as attractive. Plans which represented the older scheme of values, such as the Greater London Council redevelopment of Covent Garden or the reconstruction of the La Défense area of Paris, were bitterly attacked for just those qualities which would have made them admired a few years before. Similarly, the late 1960s saw a revulsion against motorway-building in cities, with protests as far apart as San Francisco and London, New Orleans and Paris. Earlier, it had been almost axiomatic that urban traffic should be channelled onto special segregated routes designed for the purpose. But now, opponents began to stress the disadvantages of the motorways: environmental degradation, noise, visual intrusion and severance of traditional neighbourhoods. Since it was impossible for the city ever to cater adequately for the rising tide of car traffic, objectors argued, the right policy was to restrict the use of the cars in cities and build up good public transport instead. And, by the end of the 1990s, there was an even more remarkable reversal: planners, influenced by the American 'new urbanism', began to argue that the entire notion of a hierarchical street system was wrong, and that it was best to plan urban streets for more or less equal permeability to traffic. But this was modified in turn by an even more profound and persuasive shift: traffic experts now accepted that to build one's way out of traffic congestion was self-defeating, since – after some delay – new traffic would simply be generated to fill the available new space, leaving congestion much as before.

Finally, however, the problem is that it is very difficult to reconcile different sets of values. Most planning controversies, even though the bitterness of the debate may obscure the fact, involve a conflict or right against right. Other things being equal, it would be right to build urban motorways to cater for traffic; if it can be assumed that there is no way of stopping people buying and using cars, and that in fact these cars do provide desirable personal mobility, then urban motorways are the best way of handling the resulting problem. The trouble is that this is not the only consideration. As opponents are not slow to point out, motorways are intrusive and disruptive, even if better designed than most are (which is certainly possible to achieve); funds spent on them may well be diverted away from public transport; even when the great majority of households own cars, as been the case since the mid-1950s in the United States and since the mid-1980s in Britain, the great majority of individuals at most times will still be without free access to one; and, as just noticed, the benefits will soon be swamped by further traffic growth. The controversy, then, is essentially about priorities. In a perfect world without an economic problem, there would be unlimited resources for very well-designed motorways, integrated into the urban fabric, and for equally superb public transport system available to all – not to mention all the other competing investments such as the replacement of old schools and mental hospitals and prisons, and the construction of new homes for those who are still inadequately housed. But of course the resources are far from unlimited; and the community as a whole has to decide which of many good things it wants the most.

In the final analysis, therefore, most major planning decisions are political in character. Unfortunately, as is well known, political decision-making is a highly imperfect art. Ordinary people are given the choice of voting every four or five years for a national government, and perhaps every three or four years for a local government; in either case, they must vote on a confusing bundle of different policies, in which planning issues have often been well down the list. Many of these issues, as stressed more than once in this chapter, may be so general and abstract in character that it is difficult for the ordinary citizen to appreciate their impact until critical – and perhaps irrevocable – decisions have been taken. Pressure groups may achieve effective action on particular issues, but they tend to be formed and populated disproportionately by those groups in society which are better educated, better informed and better organized – which, in most cases, also means richer. The recommendations of the Roskill Commission on London's third airport were finally overruled by the minister after a great public outcry; and many planners thought the minister right. But many also took little comfort in the fact that whereas the commission's work had cost just over £1 million, the pressure group against it spent three-quarters of a million in getting the recommendation overturned. Likewise, the inquiry in the late 1990s into Terminal 5 at London's Heathrow Airport, which took four years and cost £70 million, mainly benefited lawyers and convinced the government that a simpler way must be found, almost certainly through debates on national transport issues in Parliament. The danger here is that the greater the call for public participation in planning decisions, the greater the likelihood that decisions will go in favour of the richer and better organized – and against those who can least look after their own interests. Disillusioned, in the 1980s and 1990s some environmental activists took to the streets and even to the trees, practising direct action to stop construction of major highways like the M3 Winchester bypass or the A34 Newbury bypass. They did not succeed, but they almost certainly brought about the abandonment of other major road schemes. The critical question here was: whose interests did they represent, and in what way did their actions fit into an ordered and rational and democratic decision-making? These questions are far from being satisfactorily answered.

A particularly acute problem of divergent values, which is evident in many planning decisions, concerns the trade-off between the interests of different generations. In such situations, the best is literally the enemy of the good. Should public housing, for instance, be built to reflect the standards and aspirations of the first generation of occupiers, or the second and third? If built merely to minimal contemporary standards, the risk is that it will be regarded as substandard within a generation or two; and it may not then be possible to redesign it except at unacceptable cost. But if built in advance to satisfy the standards of tomorrow, then less resources will be available to satisfy the pressing needs of today. Similarly, many decisions about preservation and conservation involve questions of the interests of different generations. It may be cheaper to pull down a Victorian housing area in a city and replace it by new flats than to rehabilitate it; the community is then faced with a choice between the needs of those who are ill-housed, and the value of the area for generations of future citizens. Similarly, the establishment of green belts around British cities after the Second World War involved certain sacrifices on the party of those who were thereby housed farther away from their jobs in the cities, while the majority of the urban populations of that time were unable to enjoy any benefits because they lacked the cars to make excursions into the protected countryside. Here, planners may with justification claim that by their intervention they are guarding the interests of posterity, including generations yet unborn. But if fundamental values change from one generation to another – if, for instance, each generation values environmental conservation higher than its predecessor – how is that resolved? And suppose, for instance, that values vary geographically, so that unemployed people in a depressed region or town care more for job creation than the environment while rich people in a more affluent city value the environment more – how are those differences to be reflected and accommodated?

Planning in practice, however well managed, is therefore a long way from the tidy sequences of the theorists. It involves the basic difficulty, even impossibility, of predicting future events; the interaction of decisions made in different policy spheres; conflicts of values which cannot be fully resolved by rational decision or by calculation; the clash of organized pressure groups and the defence of vested interests; and the inevitable confusions that arise from the complex interrelationships between decisions at different levels and at different scales, at different points in time. The cybernetic or systems view of planning is a condition towards which planners aim; it will never become complete reality.

New planning paradigms

Because of these difficulties, as was perhaps inevitable, in the first half of the 1970s there was a major reaction against the style of systems planning – just as, a few years before, the systems planners had reacted against the master planners. In particular, some planners began to question the basic tenets of the systems approach: the notion that it was scientific, in the sense that the world could be completely understood and its future states predicted; the notion that planning could be value-free, in that the planner could disinterestedly determine what was best for society; the notion that the planner was planning for a society that was a homogeneous aggregate, in which the welfare of the entire people was to be maximized, without too much concern with distributional questions; and the notion that the task of planning was to come to terms with – which, in practice, means adapting to – the facts of rapid growth and change. These ideas had proved particularly timely in two kinds of planning, which tackled major problems of the 1960s: transportation planning, to deal with the facts of explosive car ownership; and subregional planning, to deal with the equally pressing facts of population growth and decentralization. Though subsequently criticized on technical grounds, there can be no doubt that in these fields – and in the structure plans of the early 1970s – the systems approach represented a considerable advance on the older, inflexible style of planning. In the more stagnant and constricted world of the 1970s, however, its concepts and techniques appeared to lose some of their point.

The problem, though, went deeper than that, and the attack on systems planning came earlier. First, there was the demand for public participation in planning. Beginning with official endorsement in the Skeffington Report of 1969, which resulted in a statutory requirement that participation be formally incorporated into the planning process, it struck at one of the underlying beliefs of systems planning: that of the planner as superior, scientific expert. From this, it was a short step to the notion that official participation in planning was itself a token action, designed to manipulate the public even further by offering them the shadow rather than the substance. In this view, what was needed was far more than mere consultation of the public; it was actual involvement of the citizenry in making plans for themselves. This was most appealing, but also evidently most difficult, in deprived urban areas where people were most apathetic and least well informed about the possibilities open to them.

The idea of community action in planning started in the United States, but spread rapidly to Britain in the ferment of ideas in the late 1960s, helped by the fact that at this time there was a new concern with problems of social deprivation in the inner cities of both countries. From the start, it tended towards a radical critique of society and – especially in Britain – became heavily influenced by the intellectual currents towards Marxism at the time. This was perhaps predictable: community action depended on the idea that local people should be organized, and by definition this could not be done through the agency of officialdom; the people who set themselves up in this role were almost bound to believe in some radical mission to raise the people's consciousness. In the officially sponsored community development projects between 1972 and 1977, it rapidly led to conflict between the teams and the local councils, and so to the rapid demise of the experiment. But elsewhere, in a thousand different ways, it began to generate a great variety of semi-official and unofficial groups involved in various projects, with a wide variety of political views, from liberal left to Marxist left. Many of these came to play an important role in the inner cities when, after 1977, the government released funds for partnership and programme authorities.

Marxism by then, however, was beginning to create a rather different paradigm of planning; it overlapped with community action in a number of places and in the behaviour of a number of people, but for the most part it was rather distinct. It has come to be known as the political economy approach. Its essence is this: application of Marxist theory to the development of the modern capitalist economy reveals that very striking changes are taking place in the character of the economy of the advanced industrial countries of Europe and North America, and these are in turn having strong regional and urban impacts. In particular, rationalization of production is leading to major locational shifts of industry and to big reductions in workforces, which especially affect the older, bigger, inner cities in the older industrial regions. The essence of this approach lies in analysing the changes that are occurring, and the structural changes in ownership and control that underlie them. So far the major achievements of this school lie in analysis rather than either prescription or proscription – or, to put it another way, in urban studies rather than in urban planning. Insofar as there have been policy recommendations, they tend to have been rather conventional ones in the form of an extension of the state sector, a growth of cooperative forms of production, and a control on the freedom of private industrial complexes to shut down plants. But underlying the whole analysis is a profound sense of the power of modern multinational, multi-plant corporations to affect the fortunes of cities and regions – a power that often seems far greater than the capacity of governments to influence their actions. In the event, this tradition of analysis proved pervasive during the 1980s, at a time when in both Britain and the United States right-wing governments were retreating from planning but encouraging development-led approaches to urban regeneration. And this led to a curious divorce between the theory and the practice in urban planning and development, which had never previously occurred.

The central problem with the neo-Marxist approach to planning of the 1970s and 1980s, oddly, seems similar to the problem of the systems planners whom the Marxists criticize. The burden of the Marxist critique is that the systems planners, claiming to be value-free, never realized just how value-dependent they were; they were mere technical planners, who could discuss how to reach given ends, not the ends themselves. Only Marxists, whose training has allowed them to understand the laws of human social development, could pass through this subtle veil. But once they have achieved this, presumably they – like the systems planners before them – can legitimately claim to plan and to control. The problem, for them as for the systems planners, is why anyone should heed their claim to unique wisdom? The problem as to the *legitimacy* of planning remains; and, as a progressively larger section of the public becomes interested in the impacts of planning, it becomes more acute. Whatever the planners' ideology, it appears that people are no longer willing, as once apparently they were, to accept their claim to omniscience and omnipotence.

One answer is to help an increasingly well-informed and well-organized and active population to conduct better debate. That is why much of the most interesting developments in planning theory in the 1980s and 1990s have been about a kind of transactive planning. There were beginnings of this in the 1960s, both in America and Britain, but it has now become a much more sophisticated process, informed by a good deal of philosophical underpinning that owes much to the Marxist debates of the 1970s but goes beyond them, to try to strip away levels of false understanding and false representation, and that recognizes the essential complexity of many decisions and of the machinery necessary to resolve them.

Meanwhile, just because people are conscious that planning is a public good that can have both positive and negative impacts on them, controversies over planning proposals tend, if anything, to become more vigorous and even more rancorous. In this, it is not possible to argue simply that people are fighting the planners; often it appears – as over the line of a motorway or the location of a power station – that the people are fighting each other. Especially in periods of negative growth when there are all too few goods to go around, planning may become – in the words of the American economist Lester Thurow – a 'zero-sum game': one in which if I win, you lose.

Whether zero-sum or not, few would doubt today that planning decisions critically affect what is known in the jargon as 'real income'. This, of course, is far more than money income: it includes such intangible psychic income as is provided by clean air, lack of noise, agreeable neighbours, freedom from crime, good education, a range of services accessible by efficient transport and a host of other things. One important school of urban planners, therefore, regards questions of real income distribution as central to the planning process. Plan evaluation, in their view, should be concerned less with aggregate excess of benefits over costs than with redistribution of real income as to benefit the groups that now have the least. As already suggested, in this regard aggregate cost–benefit analysis is inferior to disaggregated analysis of the Planning Balance Sheet or Goals Achievement Matrix varieties. The latter approaches have the benefit that they specifically look at the distribution of costs and benefits, and the Planning Balance Sheet specifically looks as their incidence in different groups of the population. Additionally, they are capable of incorporating elements that cannot be accurately rendered in money terms, but that nevertheless form an important element in real income, such as gains and losses in environmental quality. This was one of the important new emphases of planning in the 1970s.

Parallel to it, and owing something to the political economy school and to the distributionist school, is an emphasis on generating economic growth. This, of course, reflected the concerns of the late 1970s and early 1980s. In depressed regions such as central Scotland or Appalachia, growth was always a central concern; but by 1981 or 1982, it became all but universal. Though there were strong ideological battles between right-wing and left-wing approaches on the question – the first favouring non-planning, land-development-led approaches, the other stressing local-authority-led schemes to regenerate traditional industrial enterprise – there was an implicit agreement on the primacy of revitalizing decayed urban-industrial economies. The paradox was that at the same time, the environmental concerns of the 1970s remained strong and that, almost inevitably, they clashed with the aim of economic regeneration. The clash was starkly highlighted in the United States, where environmental groups battled with government over issues like oil exploration off the California coast, or strip mining in western mountain states; but in Britain it is illustrated in subtler ways in the arguments about urban enterprise zones, or about mining rights in national parks.

The nature of the paradox is that this is a zero-sum society, but that to get out of that state, some groups would have to sacrifice something that they hold dear. Planning, in other words, is merely an acute instance of the central problem of society in the 1980s. And in the economic revival of the middle and late 1980s, these NIMBY-style issues became more and more prominent, and local pressure groups in the more favoured areas sought to erect barriers to further growth in their areas. This is against a backdrop of ongoing dispute between rights and responsibilities of national and local levels of government and the degree to which essential projects of national importance, such as energy and transport infrastructure provision, can be 'imposed upon' local areas when opportunities for local participation, formal and informal, have increased (see Chapter 5).

Some 40 years after the Skeffington Report and 15 years of the exasperating Terminal 5 inquiry, the government finally acknowledged that some planning decisions are so political they need to be coordinated nationally and that they will affect local opportunities for public involvement. But this only opens up another (old) debate in planning – the need for greater certainty, modelling scenarios of the future, and having a credible evidence base, vital to inform critical planning decisions. This is not simply appropriate for those making decisions about major national development proposals, but also for those charged with considering future planning in individual towns, cities and rural areas.

Spatial strategic coordination

How can we conceptualize this ongoing process of planning change? Until 1997 the UK government had expected planning to be devised and implemented uniformly across all parts of Britain: one country, one system. But such an ethos appeared to run counter to sustained processes of devolution, regionalization and localization implemented since the mid-1990s. Planning has also been criticized for failing to ensure that development occurs in the right locations, or to cater for the desires and expectations of local and regional actors, and for producing standardized plans and policies that have not delivered. As a consequence of devolution, decentralization and regionalization, planning is now becoming increasingly differentiated across the UK. A uniform planning process nationally (originally devised in the aftermath of the 1939–45 war years) is incompatible with current government policies intended to foster regional economic competitiveness, sustainable communities and local distinctiveness. This relates not only to devolution for the Celtic countries and London, but also to the push for both regionalization and city-regionalization on the one hand, and addressing community and neighbourhood renewal on the other. It is just one aspect of what is expected from planning today, a type of planning we may define as a framework or coordinating mechanism for government.

This relates to a second point. A further precursor to planning reform concerns the fragmentation of the state between disparate actors and the desire for integration, with a new role for spatial-strategy making in achieving coordination between actors and their strategies. New forms of state working, governance and spatial strategies are being promoted and relied upon as tools to help resolve community, sub-regional and regional problems. This is occurring alongside what might be termed a renaissance for planning as the means of achieving policy integration and coordination and the promotion of sustainable development. In addition to addressing a lack of joined-up working between agencies of delivery in the built and natural environment, this type of planning is concerned with avoiding the danger of actor dominance and ensuring that the spaces between those fuzzy or archaic institutional boundaries do not preclude or inhibit delivery mechanisms.

Therefore, we may suggest that the emphasis here is to look at planning not as a delivery process *per se*, in the style of planning under the welfare state or in the Thatcher years, but rather as a strategic capacity and political integration mechanism intended to cement the increasingly fragmented agents of the state working within often inappropriate institutional and governmental silos. All these agents possess their own agendas, political objectives, strategies and resources, but need to cooperate in order to deliver projects and developments. Planning is being expected to ensure compatible working and strategic coordination within government, between government and citizens, and government and the market, alongside its more traditional role of land use planning within the town and country planning system. Such a responsibility has already formed a significant element in new government policy on the role of planning post-2004, and influenced the professional planning body's agenda on the role it expects its members to perform in the twenty-first century.

These institutional drivers of change within the ever-broadening activity of planning are transforming the activity and scope of planning, across scales and across territories in varied ways and at varied times; we might even say that planning is becoming plural. Planning is a contributor to and a reflection of a more fundamental reform of territorial governmental management that aims, *inter alia*, to improve integration of different forms of spatial development activity, not least the state's concern about the delivery of housing growth and economic competitiveness alongside the provision of infrastructure. The objective of this transformation is to widen the trajectory of planning, or spatial-strategy making, in the modernization and governance agendas at both the regional level and the local level within the UK. It is a process occurring across the globe at the present time.

The recent reforms to planning and the attendant regulatory and policy components have emerged as part of a wider pragmatic agenda often termed new localism. Its origins derive from pressures on government to consider whether it should divest more central power to the local level. Devolution and decentralization within the UK have enabled these principles to be tested out uniquely within each part of the country in the new relationship between the central, devolved and local states. Government has promoted an agenda of state infrastructure revitalization, decentralization and local responsiveness, cooperation and partnership with civil society together with social responsibility. Conceptually, the reformed planning system is charged with coordinating and delivering on the spatial aspects of a range of policy agendas being brought to bear at the local level, and with providing a mediation forum for various interests that is locally responsive and flexible to changing local conditions.

The question finally comes back to this: what, then, is the methodology of planning? How does it seek to resolve such a set of major problems? The answer should surely be: by some variant of the systems approach. It should not claim the instant ability to solve complex problems. It should not even necessarily claim unique expertise. It should certainly not claim to know what is good for people. Rather, it should be exploratory and instructive. It should aim to help decision makers and communities think clearly and logically about resolving their problems, and in particular some of the more subtle underlying issues that concern such matters as equity or growth. It should try to examine alternative courses of action and trace through, as far as possible, the consequences of each of these for different groups of people in different places. It should not seek to avoid the difficult questions of who exercises political power on behalf of whom, and by what legitimacy. It should make recommendations, but it should not seek to impose prescriptions. It should claim modestly that planners may perhaps be more capable than the average person to conduct this kind of analysis, but not that they are uniquely expert. In other words, it should aim to provide a resource for democratic and informed decision-making. This is all that planning can legitimately do, and all it can pretend to do. Properly understood, this is the real message of the planning experience of the last 70 years.

Further reading

The best introductory textbook on the systems view of planning is still J.B. McLoughlin, *Urban and Regional Planning: A Systems Approach* (Faber, 1969; paperback version, 1970). It should be supplemented by George Chadwick's *A Systems View of Planning* (Pergamon, 1971), which is more complex and theoretical in character. Good general books on planning, written from a systems standpoint, are Michael Batty, *Urban Modelling: Algorithms, Calibrations, Predictions* (Cambridge University Press, 1976) and David Foot, *Operational Urban Models: An Introduction* (Methuen, 1981). For a practical approach to planning in the systems tradition, see Ray Wyatt, *Intelligent Planning: Meaningful Methods for Sensitive Situations* (Unwin Hyman, 1989).

Andreas Faludi, *Planning Theory* (Pergamon, 1973; with accompanying *Readings in Planning Theory*) deals with the systems and other approaches to planning in some detail.

For the Marxist approach, see Manuel Castells, *The Urban Question* (Arnold, 1977) and David Harvey, *Social Justice and the City* (Arnold, 1973).

A fuller account of the development of theory and its relation to practice is to be found in Peter Hall, *Cities of Tomorrow: An Intellectual History of Urban Planning and Design in the Twentieth Century* (Blackwell, 1988), and in John Friedmann, *Planning in the Public Domain: From Knowledge to Action* (Princeton University Press, 1987).

For transactive planning, see John Forester, *Critical Theory, Public Policy, and Planning Practice: Toward a Critical Pragmatism* (State University of New York Press, 1993) and John Forester, *The Deliberative Practitioner: Encouraging Participatory Planning Processes* (MIT Press, 1999); Patsy Healey, *Land Use Planning and the Mediation of Urban Change: The British Planning System in Practice* (Cambridge University Press,1988); and Patsy Healey, *Collaborative Planning: Shaping Places in Fragmented Societies* (Macmillan, 1997). Critiques of various recent theoretical approaches to planning can be found in Nigel Taylor, *Urban Planning Theory Since 1945* (Sage, 1998); Philip Allmendinger and Mark Tewdwr-Jones, *Planning Futures: New Directions for Planning Theory* (Routledge, 2002); and Philip Allmendinger, *Planning Theory* (second edition, Palgrave, Basingstoke, 2009).

For the state of British planning since the 1980s, see Andy Thornley, *Urban Planning Under Thatcherism* (Routledge, 1991); Philip Allmendinger, *Planning in Postmodern Times* (Routledge, 2001); Mark Tewdwr-Jones, *The Planning Polity* (Routledge, 2002); and Gerry Stoker, *Transforming Local Governance: From Thatcher to New Labour* (Palgrave, 2004).

Index

Note: Page numbers in **bold** are for figures and plates, those in *italics* are for tables.

eBooks – at www.eBookstore.tandf.co.uk

A library at your fingertips!

eBooks are electronic versions of printed books. You can store them on your PC/laptop or browse them online.

They have advantages for anyone needing rapid access to a wide variety of published, copyright information.

eBooks can help your research by enabling you to bookmark chapters, annotate text and use instant searches to find specific words or phrases. Several eBook files would fit on even a small laptop or PDA.

NEW: Save money by eSubscribing: cheap, online access to any eBook for as long as you need it.

Annual subscription packages

We now offer special low-cost bulk subscriptions to packages of eBooks in certain subject areas. These are available to libraries or to individuals.

For more information please contact
webmaster.ebooks@tandf.co.uk

We're continually developing the eBook concept, so keep up to date by visiting the website.

www.eBookstore.tandf.co.uk